THE
BLACK THISTLE

ANOTHER LUKE TREMAYNE ADVENTURE

THE BLACK THISTLE

A SCOTTISH CONSPIRACY 1651

GEOFF QUAIFE

ARPress
ILLUMINATING IDEAS
EMPOWERING VOICES

ARPress
45 Dan Road Suite 36
Canton MA 02021

Hotline: 1(800) 220-7660
Fax: 1(855) 752-6001

Ordering Information:
Quantity sales. Special discounts are available on quantity purchases by corporations, associations, and others. For details, contact the publisher at the address above.

Printed in the United States of America.

ISBN-13: Softcover   979-8-89356-767-0
          eBook   978-1-970081-75-6
          Hardback   979-8-89389-394-6

Library of Congress Control Number: 2024916231

THE LUKE TREMAYNE ADVENTURES

(In chronological order of the events portrayed)

LEADING CHARACTERS

The English Army

Luke Tremayne	Cromwell's special agent.
Harry Lloyd	Lieutenant, Luke's deputy.
Andrew Ford	Luke's senior sergeant.
John Halliwell	Luke's sergeant.

The Scottish Army

David Burns	Leader of special unit.
Dugall Sinclair	Burn's deputy.
James Cameron	Commander of Highland infantry regiment.

Residents At Castle Clarke

Alistair Stewart	Distant relative of Charles II.
Elspeth Stewart	Alistair's wife, and daughter of Earl of Barr.
Duncan Caddell	Fanatical moralistic minister of the Kirk.
Derek Clarke	Owner of castle.
Fenella Clarke	Derek's wife.
Janet Hudson	Wealthy heiress?
Mungo Macdonald	Claims to be minister of the Kirk.
Duff Mackail	Maverick commander of Clarke's garrison.
Aiden Mackelvie	Local laird.
Malcolm Petrie	Lawyer, diplomat, scholar.
Morag Ritchie	Young girl, dispossessed heiress.
Gillian Shaw	Fenella's servant.

Other Scottish Gentry and Nobles

Earl of Barr	Scottish noble, and leading politician.
Alan Campbell	Local laird.
Jean Campbell	Alan's wife.
Lady Dalmabass	Gentlewoman accused of witchcraft.

Real Historical Figures

Charles II	crowned King of Scotland, January 1651; in Britain 1650-1; then exiled; restored to English and Scottish thrones in 1660.
Marquis of Argyle	leader of Kirk party, of the Campbell clan, and Scottish Government until early 1651; ambivalent to Charles II. Later co-operated with English military occupation, and was executed by the King on his Restoration.
Oliver Cromwell	Commander of English army in Scotland. Defeated Scots at Dunbar in September 1650, and again at Worcester a year later following Charles's invasion of England. Eff ective ruler of England, Scotland and Ireland, 1653-58 as Lord Protector.
George Monk	Former Royalist, Cromwell's leading general in the subjugation of Scotland. Eff ectively ruled Scotland for Cromwell 1654- 58. In 1660 marched army into England to restore order, and eventually the King. Became Duke of Albemarle and commander of the King's army.

HISTORICAL PROLOGUE

The Scots revolted against Charles I in 1637, and following their victory Scotland became independent of English interests for more than a decade. When the English army forced the execution of the King in January 1649 England declared itself a republic, but the Scots proclaimed Charles's son, as Charles II, King of Scotland. Nevertheless throughout 1649 and 1650 the Scottish Presbyterian government ignored the new King, and kept him powerless. The King's position was dramatically improved by two English military victories against the Scots in late 1650. These convinced the Scottish government that it could not resist the English invasion without the monarch's assistance.

Scottish independence was in peril. The King would restore an English dominated monarchy, while the English republic would absorb an insignificant Scotland into a united Britain. It was into this whirlpool of an anxious and frustrated Scottish nationalism that Cromwell plunged his agent, Luke Tremayne.

1

Oban, Scotland, February 1651

Luke Tremayne and his sergeant, Andrew Ford, disguised as disreputable locals, entered the overcrowded tavern of Widow Abercrombie. They were almost overwhelmed by a combination of human sweat, stale urine and vomit, and were relieved to immediately isolate their contact. He was sitting as pre-arranged in a corner of the room holding a blue ceramic tankard in his left hand. Luke moved towards him, while Andrew protected their retreat. Suddenly four hirsute ruffians appeared beside the contact. The tallest of the group, with an exaggerated swing of his sword, decapitated the unsuspecting drinker.

A head and broken tankard, rolled towards Luke. He changed direction, praying that the sword-wielding murderer had not noticed his interest in the recently deceased. Luke retreated, gulping down a sweet brown ale as he crashed into Andrew in his haste to leave the tavern. The environment, the murder, and the quality of the ale proved too much for Luke who just made the street before he vomited. Andrew and Luke found respite in a churchyard opposite the tavern. They sat on an elevated grave and assessed their situation. Andrew muttered that their mission was over before it began. Luke agreed that the decapitation was a major blow to their fledgling enterprise. Cromwell had ordered Luke's unit to the Western Highlands where the details of their latest clandestine operation would be revealed by a prearranged contact at Widow Abercrombie's. Andrew was uneasy. He intently watched every customer who left the alehouse and eventually commented, 'If the killer noticed our interest in his victim he could come after us.'

'Quite possibly,' replied Luke. 'Let's return to the ship immediately.'

Luke had been surprised that the ship that brought him and some sixty troopers from Dublin was the English republic's newest man of war, the fifty four-gun *Providence*. The warship remained anchored in the sea loch while Luke and Andrew came ashore to receive details of their mission. The two soldiers left the churchyard in growing darkness and light sleet. They made their way along the foreshore heading for their concealed dinghy. Gusts of snow drove into their faces adding to their discomfort as they removed the branches of broom, which covered the dinghy. This discomfort turned to apprehension when several swordsmen emerged from the gloom and surrounded them.

Their leader spoke, 'Do not be alarmed Colonel Tremayne. I am your contact, Major David Burns, acting for the Scots government under specific orders from the Earl of Barr, and with the approval of General Cromwell.'

'Then who was decapitated at Abercrombie's?' asked a suspicious Andrew.

'A local vagabond. I persuaded him to take my place for an evening of free drinks, and the use of my large blue coloured tankard. I planned to watch developments from a safe distance.'

'How do I know that you are my contact?' asked a sceptical Luke.

'You don't. But without me you don't have a mission.'

'So what then is my mission?' demanded an annoyed Luke.

'No discussion in the middle of a snow storm,' shouted the Scot. 'Return with me to the town, and I will put you in the picture.' Andrew looked quizzically at Luke. David Burns appeared convincing, and the five men who accompanied him gave the Englishmen little choice.

After a satisfying meal of chicken and leek soup, Scotch pies and oatcakes Luke and Andrew settled in for a long session of whisky drinking as David explained the situation. It gave Luke an opportunity to examine his new ally. David Burns was tall and well built. He had dark brown eyes, black hair that just reached his shoulders, bushy black eyebrows, and a small well groomed black beard. His face was long, with a protruding chin. He exuded confidence and charm, and as he spoke he revealed a passion for his cause, a passion controlled by a decade of experience as a cavalry officer.

As he spoke he removed more of his outer garments revealing a deep blue doublet, a surprisingly large lace collar and equally large lace cuffs. This was a man of considerable wealth.

Luke paid enough attention to David's briefing to understand that the Scots were divided into three political groupings–the Episcopalian and Catholic Royalists who had been excluded from government for years; the moderate Presbyterians who accepted the necessity of reconciliation with the so-called King but differed on just how much power Charles Stuart should have; and the Presbyterian hardliners who accepted the monarchy in principle but had little time for young Charles. Anticipating difficulties Luke asked, 'Where does our mission sit with regard to these three factions?'

David answered obliquely, 'Over the last few months pressure has been placed on many political notables and their families to change their political position in favour of the King. People have been abducted, and some even murdered.'

'That's not unusual in this country,' commented Andrew. David ignored the comment. 'A few weeks ago several families looking for their missing relatives discovered independently they were being held in a castle just a few miles up the coast from here.'

'But why is this important to Cromwell?' asked Luke.

'A fatally wounded Scottish trooper was found by an English patrol. Before he died he revealed that a major plot was being hatched against Scottish leaders and English generals within Clarke Castle–the same castle where most of the kidnapped Scots are being held.'

'Come, Cromwell would not mount a secret expedition on such flimsy evidence–a dying man claiming there is a plot to kill English officers. There are a dozen such plots each week against the General alone,' replied a suspicious Luke.

'You are right Luke. The trooper's evidence only confirmed what had been reported independently to the Earl of Barr, who with the Marquis of Argyle has ruled Scotland for a decade. It is known that the King has set up a clandestine group to wreck vengeance against those who murdered his father. These targets include many Scottish aristocrats and English officers, the most prominent of which is Oliver Cromwell. The Earl immediately informed General Cromwell.'

'Why would Barr betray his King to the English invaders? Isn't that treason?'

David smiled, 'Don't be naïve, Luke. One option available to the current Scottish political leadership is to reject the King, and ally with the English invader. The Earl is keeping his options open.'

'But it is not likely,' said Andrew. 'Barr hates the English. His family protected the Borders against us for centuries.'

The discussion was interrupted by shouting and the clash of steel on the landing outside Burns's chamber. Burns's men held back a push of six assailants who pressed up the stairs. As additional defenders emerged from the room two of the attackers fired their pistols. The shots infuriated Luke. He swung his sabre with maniacal frenzy that had the enemy reeling backwards and tumbling against each other in obvious disarray. Luke reached one of the pistoleers, who disconcerted by his falling colleague, failed to fend off a slight blow to the neck. Andrew fired his carbine at the retreating men with only partial success. One man fell, rolled to the bottom of the stairs, staggered to his feet, and fled after his comrades.

Back in the chamber Luke dressed David's superficial wounds and asked, 'Were they after you? Is this some long standing local feud?'

'No, they were after you, not us. They were the same group that decapitated the vagrant at Widow Abercrombie's,' replied David. One of his men re-entered the room and handed David a pouch which he claimed he found at the bottom of the stairs. David extracted a folded paper from it, and eventually handed it to Luke. It was a coded message signed 'The Black Thistle.'

Before Luke could express his disappointment David indicated that he could decipher the message, 'It's a code used by the Scottish Government. The letter instructs the recipient to kill you and your contact before you leave Oban. If that fails the assassins are not to follow you, as other agents at Castle Clarke will complete the task.'

'And who is this Black Thistle?' asked Luke.

'It's the organization I referred to earlier, set up by the King to complete tasks that he does not want to be associated with. Part of your mission is to uncover its local leadership.'

'What can you tell me about it?'

'After the defeat at Dunbar at least two members of the Committee of Estates, the body that rules Scotland, received letters encouraging them to accept the King unconditionally, and not to assist the English. Since your General Lambert defeated the Western army of Kirk extremists some weeks ago, their fortunes and that of the current Scottish Government have slumped. Several of their powerful supporters have since been murdered by the Black Thistle.'

'This makes no sense David. Why would the King bother? Everything is going his way. When Lambert defeated the Western Army these fanatical supporters of the Presbyterian government lost a powerful military force that could have been used to balance the rising power of the King within the official Scottish army. More importantly they have simultaneously lost their majority in Parliament, and in the General Assembly of the Kirk. The areas that gave them their strongest support are now under English military occupation. They cannot send their representatives either to the Scottish Parliament or assemblies of the Kirk. This now gives those more favourable to the King a majority. Most Scots would agree with the Black Thistle that their only hope of fending us off is to give the King real control of the army, and strengthen it with thousands of currently debarred Royalists.'

'Yes, you could be right. The King's fortunes have risen in recent weeks, and he no longer needs the Black Thistle.'

Andrew interrupted, 'Maybe the Black Thistle has taken on a life of its own and is acting independently of the King.'

'I don't know, but our mission remains the same. Rescue the prisoners at Castle Clarke, and uncover what exactly the Black Thistle is currently about,' responded David.

'Well so much for the strategy, what are our tactics?' asked Luke.

'Take Clarke Castle, and interrogate the prisoners and their keepers.'

'If Clarke Castle is such an item of interest, why does it need English troopers to capture it? Why haven't you taken it yourself ?'

'The castle is on a rocky promontory. My men have invested it for weeks but it has thick walls and can only be taken if these are breached from the sea. We lack ships and cannon.'

Luke now understood. 'I see why my men were transported here in the latest and most powerful of English warships. It is not my troop, but the *Providence* that you need.'

'Yes, the deal my master made with Cromwell was that he would assist with the capture of the castle, and we would share the intelligence. As there are factions on both sides that would not agree with this co-operation our masters chose clandestine units such as we both lead to carry out the operation. If things go wrong our respective masters will deny any knowledge of this enterprise.'

Luke and Andrew rejoined the *Providence* and sent the ship's pinnace back to collect David and his men. Andrew was not happy.

'I do not trust Burns. We have no evidence that he is the contact that Cromwell indicated you would meet.'

'In the circumstances we cannot expect definitive proof. Written instructions would endanger the security of the joint mission. The General would not have committed so many men, and *The Providence*, unless he was convinced of the genuineness of the mission, even if we cannot completely vouch for David.'

'But there is more,' claimed Andrew. 'If Burns's unit is the elite group he claims it to be, how did all the assailants escape? Did you notice that none of Burns's men were seriously injured? And how convenient it was to find an important document at the bottom of the stairs. His own men probably put it there.'

Next morning aboard the *Providence* Luke brought his own troopers up to date, and pleaded with David to convince them of his credentials. He responded in generalities. 'I fought with you at Marston Moor six years ago when both of us believed the Scottish and English Parliaments could together force Charles I to govern in support of the true church. Your execution of the King turned virtually all Scots against you, and now the key question in Scotland is on what terms Charles II will be supported. He was crowned King of Scotland a month ago and will lead the Scottish army against you in the summer.'

'Your master, the Earl of Barr, leads the current Kirk controlled government, but is under heavy pressure to include the King. What will Barr do?' asked Harry.

'The Earl of Barr took up arms against the boy's father to preserve Scotland's religious freedom, and now after over a decade of Scottish independence he does not want to lose it—either to an English military conquest by the current republican government, or to a restored monarchy based in London. The Kirk party is split over the King's desire to invade England, and be restored to his English throne. The more nationalist of Presbyterian royalists want Charles as a powerful King of Scotland, but want him to renounce any link with England. They want to recreate the independent Scottish monarchy.'

'Wasn't that the policy of the defeated Presbyterian extremists who formed the Western army?'

'In part, but those fanatics ignored the King. The Earl of Barr recognizes that the Scots want the King back in control, and hopes to divert his English ambitions and maintain Scottish independence by political means.'

Andrew and Harry remained unconvinced. David was a smooth talker and an experienced soldier but he was not someone they could trust. If they had seen him at that very moment their distrust would have been validated. As David moved away from the English soldiers he smiled as he thought how rapidly events had moved—and how unpredictably. A few weeks earlier he had been with the King of Scotland.

2

Stirling Castle, Seven Weeks Earlier

David Burns stood before the Earl of Barr, a small man with thinning, closely cropped grey hair. He was dressed totally in black, with a small lace collar and equally small cuffs. His black clothing, highlighted by extensive use of gold and silver thread, reflected the conflict within the man— from a distance the dour dedicated Presbyterian, up close a powerful and wealthy nobleman. His large head and bulging eyes and slight hunchback reminded David of the dwarfs that used to frequent the Scottish Court.

But Barr was no laughing matter. His family had protected Scotland's borders with England for centuries. They had manned key towers along the frontier. They were fiercely Scottish, and detested the English. In 1603 James VI of Scotland became King of England and disarmed both sides of the border. Defences were demolished. The unhappy protectors of Scotland's territory were offered a new and unappealing life as peaceful farmers. Many families failed to adapt, and the King relocated them to Ireland. The most recalcitrant families were outlawed. Their family name was never to be spoken, and its members were dispersed amongst other clans.

The one Borders nobleman to adapt to the new circumstances, and actively enforce the King's new policy was the father of the current Earl of Barr. His support of James, now settled in London, reaped him great rewards in lands, tenants and political power. While the father gained power and wealth by supporting the absent King in the first two decades of the century, his son increased that power in opposition to James's son Charles in the sixteen thirties when the King tried unsuccessfully to impose a religious change on Scotland. With the Marquis of Argyle the present Earl of Barr led the Presbyterian nobility in opposition to Charles and the two nobles had effectively ruled an independent Scotland for over a decade.

David realised that something was amiss. The room was heavily guarded. Barr looked pale and somewhat distraught. There was a distinct smell of gunpowder and charred wood. David had been surprised as he rode into Stirling Castle that while the outer defences were in the hands of long serving regiments whose artillery prowess had been proven time and time again, the inner courtyard's defences were split between the Earl of Barr's men and a motley collection of hastily recruited Irish Royalists.

'My lord, what has happened?' asked David.

'I was making my way along the corridor when two or more grenades were thrown at me. Fortunately I passed behind a large pillar at the critical time. The column that saved my life is pitted with deep gouges.'

'Is that why you moved your own regiment from the plains beneath the western wall into this building?'

'No, I did that a week ago. I was concerned by the growing strength of the royalist enclave within the Castle, and the rogues and ruffians who claimed to be the King's bodyguard. I tried to persuade the Parliament to impose a Royal bodyguard officered and manned by loyal Presbyterian Scots, but the King refused to co-operate.'

David had never seen his patron and commanding officer so anxious. The Earl had taken an early interest in David, perhaps because of his troubled family background, and eventually rewarded him with a key military position, and marriage to his niece. David in turn had been a loyal and devoted servant who admired Barr's skills, and his determination to maintain Scottish independence. His only disagreement with the Earl had been a personal matter over which he still remained aggrieved. But this was not the time for personal issues. He asked, 'My lord, why am I summoned here on this bitter winter's morning?'

'The defeat of an effective army of God fearing Scots, the Western Army, is the trigger for my summons. That defeat has led a majority of my colleagues to turn their back on the past, and to rally to the King. They fear that without Charles and his potential army of catholic Highlanders, Irish mercenaries and various assorted ruffians and profligate sinners, we will not be able to resist the English.'

'How can I help?' asked David.

'Charles and his entourage have taken up residence in the Palace precincts of the Castle, just across the courtyard. There is a constant flow of notables visiting that wing, and many Royalist leaders, debarred from service in the military or of occupying positions of power in the state, nevertheless appear to participate in a rival governmental and military structure. My secretary tells me to flee while I can. Charles's growing popularity will soon enable him to take over the official army and the state. My opposition to his father, and my current reluctance to place the government of Scotland in his hands will not be forgotten.'

'Sadly my lord I agree with that assessment. To preserve Scottish independence from the English military republic we now need the help of the Scottish royalists. Parliament and Kirk are ready to repeal all the restrictions against them. The King does not need to act illegally against anybody. He will be invited to lead the army and the state in a matter of days, and you and Argyle will be quietly sent to your estates. The King would be a fool to act against you personally because you still command massive support in parts of the country. You need each other if Scotland is to be saved. Consequently I do not think this grenade attack was a Royalist plot,' concluded David.

The Earl simply grimaced. David asked, 'What precisely do you want me to do?'

'In the future Charles will not need to act illegally, but in the last few months because of his lack of real power he embarked on several illegal and immoral activities that he now wishes to conceal. Just as your unit acts for me in clandestine operations young Charles had a group of Royalist fanatics, the Black Thistle, carrying out his less reputable activities. He meets with them on a regular basis. Every afternoon since Charles arrived in Stirling he rides to the nearest woodland just beyond the Castle to hunt. He takes only two or three friends, including my son-in-law, Sir Alistair Stewart. My spies reveal that every last Wednesday in the month he leaves his companions to hunt by themselves, and alone rides deeper into the woods. He returns an hour or so later, but without any game. I suspect he meets the Black Thistle or some other group that we must monitor. Today is a relevant Wednesday. I want you to follow him.'

David left the chamber uncertain as to where he stood in the rapidly changing political situation. He was a Scottish nationalist who strongly supported the independent policies pursued by Argyle and Barr for over a decade. England had been too busy with its conflict with King to dominate the Scots. Now the English Republic seemed determined to bring Scotland under its control because the Scots had proclaimed Charles's son as Charles II who in turn was determined to regain his English throne. As he left across the inner close David was jeered at by the Palace guards who were Irish mercenaries in the service of his King. David reached the outer defences where his own men had waited. He ordered all but two of them back into the Castle to join the rest of the Earl of Barr's regiment. He then headed with two remaining troopers for the woods.

They hid off a small clearing where Barr's agents claimed Charles split from his companions. An hour later David heard the chatter and laughter of a few horsemen making their way through the trees. Charles was indistinguishable from his companions. The Earl had been right. One of the King's companions was David's own cousin by marriage, Sir Alistair Stewart. Alistair had married the Earl of Barr's only daughter, while David had married the Earl's niece. Alistair's presence was not surprising. His family had always been fervent Royalists. When the English executed the old King, the Earl of Barr with forethought felt a family alliance with the new King might hold him in good stead. He had not foreseen that they would be in direct competition for control of the Scottish state so soon. David blamed the English victories for this unexpected development. They had destroyed Barr's power.

The influence of the Kirk, and Charles's lack of money prevented any royal display of conspicuous expenditure. For young Charles hunting remained a private relaxation rather than a public display. After a few gulps from his flask Charles waved his companions farewell and trotted further into the forest. David followed on foot. A noisy horse would give him away. His men remained to watch the King's companions. Charles also immediately dismounted, and led his horse into the ever-darkening forest. The snow-laden conifers were so thick that combined with a clouded winter afternoon, light was almost excluded from the woods. In the relative darkness David easily saw the flickering of candlelight emerging from a hut. Charles entered this small building which David circumnavigated, noting that it had one door, one window high in the wall opposite the door, and a low additional building attached to the side.

On the far side of the hut there were eight or nine horses with five or six attendants. A large white stallion munched contentedly on some vegetation that protruded from the snow. There were three smaller spotted ponies that David did not recognise, and as many heavier brown cavalry mounts.

The attendants were unhappy with the cold wet conditions and were struggling to ignite a pile of dampish sticks and twigs. They left their equine charges loosely tethered on the outskirts of a comradely circle they had created. One of them, noticing the smoke cascading from the chimney of the hut, entered the building and emerged soon after with a bucket of glowing coals. They were so engrossed in getting warm that David crawled around the edge of the circle undetected, untying the horses and encouraging them to wander off into the forest gloom. David felt quite satisfied with his disruptive activity.

As he eased himself around the perimeter of the hut to its far side he heard the squelch of a heavy weight on the snow. Before he could identify the intruder he was knocked to the ground by the impact of a spear thrown from a short distance. A dazed David lay motionless. The squelches came closer. David struggled to assess the extent of his injury as a tall figure bent over him, with dirk drawn and fatal intent.

3

The assailant came too close and momentarily relaxed. He bent over the prone body to examine the effect of the spear. David grabbed his wrist, knocked the dirk from his hand and managed to push his own dagger deep into the attacker's chest as he pulled him to the ground. The snow turned bright red as the man bled quickly to death. David removed the spear. It was a light-hunting javelin which had not penetrated his breast place but had slid under his armpit where it had done little harm. David dug into the snow with his hands, and rolled the body into the cavity he had created. He covered the grave with some branches and tried to conceal the pool of blood with additional snow. Despite his attempt, and the continuing fall of snow a circle of pink remained visible even in the gloom. He then heard the chatter of the men attending the horses coming closer.

Two men came round the corner of the hut as David pressed himself against its wall. One of them called for their missing comrade several times. The second man intervened, 'Don't waste your voice. He has gone off on his own hunting expedition. He had his javelin with him. Let's find our horses before they wander too far.' They rambled away treading on the pink stain that for the moment dominated David's world. He held his breath.

Regaining his composure David pushed open the door of the attached outhouse. It was half stacked with wood. The door on its far side opened into the hut itself. With his ear pressed against this inner door he listened. After some time he identified four distinct voices. A boyish voice proclaimed. 'In a matter of days I will truly be King of Scotland. I will not meet you again, and will absolutely disown your activities if they are revealed.'

13

'We understand, but how will we receive our orders?' asked a voice with the hint of a Highland accent.

'I will shortly appoint a new leader of The Black Thistle. He will pass my orders onto you.'

'How will we know that these orders are genuine?' asked a high- pitched voice.

'The new master of the Black Thistle will wear this amethyst ring of mine as a symbol and validation of his position. Now what have you to report?'

It was the cultured voice whom David deduced was the leader of the group who responded, 'Sire, we are well advanced in eliminating those Scottish and English leaders who participated in the murder of your father. So far these deaths have been seen as part of the clan feuds and family disagreements so common in our part of Scotland. Our secondary mission to encourage people to join your cause is no longer paramount. The English have done our work for us. The Presbyterian government can no longer maintain Scottish power without your help.'

'Don't be too sure of that,' interjected the highland accent. 'Our countrymen have another option which sadly some will follow, unless we keep up the campaign of terror against them. They will support the English in hope of future advancement should the English military invasion be successful. Throughout centuries of our history there have always been clans and families that have supported the English against a legitimate Scottish King.'

'You forget gentlemen that I am also the legitimate English King. I cannot see many Scots siding with the republican rabble against their traditional monarch.'

'Sire, over the last few weeks we have removed several lairds who supported the Western Army, and several of those close to Barr,' continued the cultivated voice.

'Do what you can to advance my cause, but only eliminate those I have named,' replied Charles. 'God will forgive me. These men murdered my sacred father. I am God's avenging angel. When I ascend the throne of England I will forgive all my father's enemies, except those that murdered him.'

'Then sire you must be very happy with the current turn of events,' whimpered the guest with the high-pitched voice.

'In general yes, but there are some problems stemming from your activities in the West. That is why I wanted to speak to the three of you personally, and not just your leader. In following your personal vendettas and putting pressure on my political enemies by abducting them or their closest relatives, you have unknowingly upset several of my closest supporters. For example Barr's daughter is also wife of my kinsman and friend, Sir Alistair Stewart. No harm must come to her. In the changing political circumstances you must release all of the women you have kidnapped. Is that clear?'

'For the moment sire their continued incarceration may help us get closer to your remaining enemies. They remain bait for our main mission,' firmly answered the highland accent. He continued, 'It is getting cold I will stoke up the fire.'

David froze. Footsteps approached the wood chest. He scrambled to get outside before the inner door was opened. The outside door was jammed. He was trapped. He clutched his sword as he awaited discovery. The cultivated voice shouted, 'Don't use the wood in the hutch. The woodsman said it was green. The wood stacked inside this chamber is the drier.'

David relaxed. Charles continued, 'When I am King all your illegalities will be wiped clean. Should you die in the cause you will, like my father, be martyrs.'

The cultured voice reported further, 'For the moment your cause progresses well in the Western Highlands. I am unassailable in my castle, although I believe the Scottish government is sending an army against me.'

David smiled. He had a location, individuals and a group to follow. Charles asked, 'How will you gain access to the English military headquarters?'

The cultured voice answered, 'To achieve it we may have to appear as traitors to your highness. We will rely on English gullibility and arrogance.'

'We all have to dissemble,' answered Charles.

David had heard enough, but he was still unable to open the door of the wood hutch. He had to stay until the conspirators left. When he thought they had all gone David took a small log and with as much force that he could muster in such cramped conditions, smashed it against the latch. The door splintered rather than opened, and he crawled out into the room. The fire was still blazing and it provided the light that the faltering and spent candles failed to give. David had felt the cold in his confined space and was warming himself in front of fire when the door burst open. A burly woodsman, with his axe at the ready, strode into the hut.

He hesitated. He did not know how to react when confronted by a senior officer, and clearly a social superior. David sensed his opportunity. 'The gentlemen, and the very special gentleman have just left. I have to ensure that they leave no evidence of their presence, or their identity, before I can hand back your hut.' The woodsman seem satisfied and suggested that they should share the whisky that the gentlemen had left behind. David took a couple of swigs but indicated that he had to catch up with the others.

David walked back to collect his horse in the gloom. Suddenly he heard shooting–multiple musket shots. He ran through the forest as his men rode towards him with his horse. The King's party was being attacked and were held down behind a fallen log. The King and his men only had pistols, swords and their hunting spears. David ordered his men to spread out and make a lot of noise as if directing a multitude of horsemen. The potential assassins withdraw on hearing the din believing that they were heavily outnumbered. David directly confronted one retreating assailant and shot at his hooded but unprotected face. The pistol misfired, singeing David's own hand. Within minutes the attackers had disappeared. David rode up to the King's party. A suspicious Charles asked, 'And who are you sir?'

'I am Major Burns of the Earl of Barr's Regiment. I am on garrison duty at Stirling Castle. My men and I were on a regular patrol of the woods when we heard the shooting.' Alistair acknowledged his cousin with a nod, and informed the King of their relationship.

'Thank you for your timely intervention Major. You can escort us back to the castle.'

The King's companions looked pale, but resolute. By their speech they were all Scots. David asked, 'Sire, do you know the men who attacked you?'

'No Major, as you saw they were all hooded.'

David lied, 'My men and I saw at least three men on the other side of the woodcutter's hut. Your attackers might be these men.'

Charles looked troubled, 'You know the woodcutter's hut?' David continued his lies; 'On our patrols we often pass the hut, and exchange greetings with the woodcutter.'

'Did you see him today?' asked an anxious Charles.

'Yes, he was cutting wood well away from his hut,' David replied.

Charles relaxed. They continued in silence. David was troubled. The man who led the attack on the King rode a white stallion and his order for his men to withdraw was delivered in a high-pitched voice. A member of the Black Thistle fraternity was the would-be assassin. The King had a traitor in his midst.

The group reached Stirling Castle. Charles asked David to enter the Palace, and drink with him. It was next morning before David was free to report to the Earl of Barr. The Earl sat expressionless as David revealed a selective version of what he had discovered. He made no mention of the white stallion and the high-pitched voice. He was disconcerted when the Earl pressed him on those very points,

'Did the attackers identify themselves in any way—their voices, their language, their horses?' Barr stared intently into David's eyes willing a positive answer. It forced an almost involuntary response, 'It was dark and the forest thick but one of the attackers rode a light coloured horse,' admitted a troubled David.

Barr breathed deeply, took up his pen and wrote. He imposed his seal on the letter and handed it to David. 'Major, take this letter to English military headquarters in Edinburgh. I have addressed it to General Cromwell. Ensure that only he gets it. I have asked for a reply.'

'In case I have to destroy this letter before I deliver it, what does it say?'

'I am passing on your intelligence, and that of others, that a plot is being hatched in the Western Highlands to assassinate the English generals and some Scottish nobles. The English are about to build up their forces in that area, whereas the only troops we have there are Sir James Cameron's Highlanders who in the current circumstances could desert us for the King at any time.'

'Is this suggesting an alliance with the English against the King?' asked a cautious David.

'Not at all Major, but we may need English help to maintain law and order in the Western Highlands until we reach some agreement with the King. However as you imply not all our companions would understand this type of co-operation. You will be acting on your own. If our enemies uncover what is happening I will denounce you.'

David was a troubled man as he left the Earl. It was not his mission that perplexed him, but the attack on the King. The King's secret organization was already infiltrated by an enemy. Who was he? Should he inform the King of his suspicions? Should he tell Barr all he knew? Had Barr told him everything? Why was Barr so anxious to discover whether the assassins had left any evidence as to their identity? And why did he immediately write a letter to the English. Was Barr behind the assassination attempt, perhaps in alliance with the English? Was Barr a traitor?

4

The Palace, Stirling Castle, Three Weeks Later

Sir Alistair Stewart and the young King had become close friends. They had just returned from Scone where Charles had been crowned King of Scotland and had undergone a dramatic personal and political transformation. He listened less to his elderly English advisers, and more to Scots of his own age, especially Alistair Stewart. Alistair was a high-minded young man who lectured the King as to his responsibilities. The King must take the moral high ground if he were to win over the Scottish political nation. The Lowlands had almost a century of Presbyterian influences. Whereas Puritanism in England was an unpopular view imposed by the army, Presbyterianism was part of what it meant to be Scottish, and its moral imperative was an essential part of its political profile.

'And how do I begin to achieve this worthy end?' asked a partly cynical Charles.

'Abolish the Black Thistle, and denounce its policies,' replied Alistair.

'Would I not be a hypocrite to do that?'

'No. If you are honest the nation will accept it. At first you wanted revenge for your father's murder. You saw yourself initially as God's avenging angel, but after listening to the leaders of the Kirk you realise that God will take His own vengeance, and your initial attitude was not in keeping with the dignity and sanctity of Scottish Kingship.'

'Alistair, your negative attitude to the Black Thistle is influenced by its abduction of your wife. That was not on my orders. The members act on their own judgment within general policy aims. They could be acting from personal spite or simply to put pressure on the Earl of Barr. You, her husband may be my best friend, but her father is our current enemy who prevents me taking real power in Scotland by his manipulation of Kirk and parliament.'

'That's another reason for you to abolish the Black Thistle, and break all your connections to it. Not only are individual members using your authority to commit all sorts of atrocities, but I believe it has been infiltrated by your political enemies.'

'What do you mean?'

'The attack on your life after your last meeting with its Western Highland cell.'

'How does that reflect on the security of the Black Thistle?'

'I have been investigating that attack over the last two weeks.'

'And what have you uncovered?'

'A forester reported that three horsemen entered the woods from the far side on that afternoon and one of them rode a white horse. I then questioned the woodcutter thoroughly and he said that one of the horses tethered near your meeting place was a white stallion. A member of the Black Thistle that you met with that afternoon led the assault on your life later that day.'

'I know,' said the King quietly.

Alistair was lost for words. 'How did you know?'

'One member of that group spoke with a very high-pitched voice, more like a woman than a man. I heard a similar voice order the retreat after Barr's man, your cousin, arrived and saved us.'

'What are you going to do about it?'

'What do you suggest Alistair?'

'Order the new leader of the Black Thistle to implement your new policy. Dissolve the organization, and deal with this traitor through the courts.'

'My new leader is on an unexpected mission, and will not be in a position to act as speedily as you can. I will inform him of what I am doing.'

'What am I to do?'

'The leader of the Western Highland group is Sir Derek Clarke who owns Clarke Castle where he has incarcerated a number of prisoners, including your wife. Rescue your wife, and inform Clarke of my new policy. He is to release all prisoners, and place the most important under my immediate protection. He is to cease the elimination of my political enemies, and instead bring them before the courts. You will uncover the identity of the high-pitched speaking, white stallion owning traitor, and bring him to me.'

'I will leave immediately.'

'It will not be easy. My agents tell me that a joint English- Scottish force has been sent to capture the Castle by force, and free the prisoners. It will be a coup for Barr, and buy him credit with both the Scottish political nation, and the English.'

'He won't succeed. Castle Clarke is only vulnerable from the sea and with heavy naval artillery. Barr has no ships,' replied Alistair.

'But the English do. My Irish friends claim at least two new heavily armed warships are in Dublin ready to sail north. Barr is no fool. He is calling on the marine firepower of the English to subdue Castle Clarke.'

'To be fair to my father-in-law he would do anything to rescue his daughter. I doubt that any secret alliance with the English to utilize their ships reflects any betrayal of Scotland. His family have hated the English for centuries.'

Alistair left Stirling for the West Coast through glens and high passes covered with snow, and in the face of westerly blizzards. This mountainous trek would give the King's enemies less opportunity to discover and frustrate his mission. This lonely bitterly cold journey gave Alistair ample time to worry about his wife. He was particularly concerned that the daughter of the country's most important politician could be kidnapped, and nobody seemed greatly alarmed. Alistair had discussed his wife's fate with her father who simply indicated that now he knew where she was he was certain that no harm would befall her.

At least the King wanted her released. Alistair could not help but think that the source and reason for his wife's abduction might not be as clear as some suspected. What would happen when his wife delivered the child she was carrying? Control of that heir gave the guardian immense power and property. His father-in- law probably regretted marrying his daughter to a Stewart, but it would be irrelevant if Barr controlled the upbringing of his grandson. Daughter and son-in-law would be obsolete. Alistair's first priority, despite his loyalty to the King, had to be the protection of his wife and imminent child—at any cost.

The blizzard worsened and visibility was reduced to a few feet. Alistair wandered off the trail and fell into a deep pocket of snow. He was freezing and could not feel his fingers or his feet. With such short days Alistair knew it would soon be dark. He fondly remembered his childhood where Highland branches of the family looked after him. He remembered that the warmest place in such an environment was under the snow. He dug himself a snow cave with his cavalry shovel–sufficient to also provide partial protection for his horse. There was no opportunity to light a fire. He took out a couple of hard oatcakes and broke them into pieces. He used his over cloak as a mattress, and in the complete darkness and silence he was soon asleep.

The following days became easier as he descended through low passes into the wide glens of the western side of the mountain range. He saw many lochs ahead of him, and in the far distance, the sea. He was alarmed by the amount of smoke arising from the next glen. The explanation was soon clear. The glen contained a large military force. As Alistair entered the valley a group of Highlanders in long plaid cloaks and blue berets demanded his identity and mission.

'Sir Alistair Stewart on a mission for the King,' he replied, hoping that these were Royalist troops and not some maverick mob.

The men signalled for Alistair to follow them. On entering the largest hut in the hamlet Alistair appreciated the blazing fire. He recognised the man standing in front of it offering him a mug of strong drink. The officer introduced himself; 'Sir James Cameron, colonel of a regiment of Cameron Highlanders in the service of the Scottish Government.'

'That Sir James begs the question. Which Scottish Government?'

'There is only one Scottish Government–that emanating from Stirling Castle led by the Marquis of Argyle and the Earl of Barr acting in the name of the King,' was the highlander's diplomatic reply.

'It is only a matter of days Sir James, and Parliament and Kirk will give Charles their blessing and force the nobles to hand over to him executive power. And then what do you Sir James?' probed Alistair.

'No problems. I can see that the King has not kept you up to date with his war plans. He is expecting reinforcements of Irish Royalists to arrive in Spanish ships at any time. They will be unloaded on the shores of the many lochs that are part of Cameron territory, and it will be my responsibility to organize them, and then march to Stirling to join the other Royal armies that will gather soon to evict the English. At the moment I am engaged in a law and order exercise for the current government. This gives me free reign to move around the Highlands at will. I am hunting down criminals and brigands, who swoop onto the Lowlands, steal cattle, raze crops and kill innocent people. They then escape into the Highlands. What exactly is your mission?'

'I am heading for Castle Clarke. The King is not happy with what is happening there, and I am to deliver new orders to the Castle inhabitants.'

'I would be amazed if those criminals in Castle Clarke listen to the King. In due course I will have to deal with them. You are heading for trouble.'

Sir James Cameron was an impressive figure. He was a giant of a man well over six feet tall whose physical strength as a young man was renowned throughout the Highlands. He had a rectangular head with a very broad forehead, green eyes, a reddish curved nose and large mouth. These features were all jammed into the lower part of his face. His reddish brown hair was worn long in the Royalist manner, and his facial hairs were manifest in a woolly looking moustache and a short but broad beard that were more ginger than the hair on his head.

His political and military reputation was equally impressive as his personal appearance and character. He had served both the Scottish government and the previous King but managed to avoid the fatal calamities of unthinking Royalist activity. He had conserved his army for the more momentous occasions that might lie ahead. Barr could trust him to police the Western Highlands while being well aware that when the time came Sir James would rally to the young King.

Alistair was alarmed by Sir James's description of the owners of Castle Clarke but after a refreshing meal continued his journey. He eventually reached a large clearing not far from the castle. There to his amazement was a large encampment of troops. He had seen them for previous months just across the courtyard at Stirling castle. They were companies of the Earl of Barr's regiment. They were in siege formation, although Alistair could see no artillery. The regiment was basically infantry with a few squadrons of dragoons. Maybe be his cousin David Burns was in command.

In the circumstances Barr must not know that the King's personal envoy was attempting to reach the castle. Alistair would hide until he found an opportunity to enter the castle undetected. It came sooner than expected. A wagon left the castle's front gate drawn by a very large horse. The besieging troops made no attempt to rush the open gate, and ignored the two woodcutters who led the horse into the forest. Alistair followed them and when well away from the castle he approached the men, who instinctively drew their daggers. Alistair introduced himself as a messenger from the King who had urgent business with Sir Derek, and did not want to be seen by the Earl of Barr's men.

The younger woodcutter suggested that Alistair could lie in their wagon hidden under a number of logs. His older comrade shook his head in disagreement. 'Don't you know the story of the wooden horse lad? This man might be an assassin sent by the English to kill Sir Derek.'

'In that case we could arrest him and take him to the castle as a prisoner,' suggested the young woodsman.

'If you could do that without the officers of the regiment seeing me, then do it. But I must not be recognized by any of the besieging troops. Who is their commander?'

The older man replied, 'The force is commanded by a Major Burns who warned Sir Derek today that he was off to Oban, and that when he returned he would have a secret weapon capable of subduing the castle. His deputy who is currently in command is a Dugall Sinclair, 'I know both men,' said Sir Alistair.

'In that case stay hidden here. We will inform Sir Derek. He will send someone to bring you into the castle unseen,' promised the older man.

The woodsmen chopped and stacked for several hours and eventually moved back to the castle. Alistair was feeling cold and hungry as he hid under a large bush. Light drizzle was falling, but the tree canopy kept most of it away. Several hours into the night Alistair was alerted by the sound of a bird. The sound was coming nearer. Alistair made a pathetic effort to mimic it. Immediately after which he heard a laugh and made out the figure of a short but athletic man approaching him. The stranger asked whether Alistair was the envoy from the King, and introduced himself as Duff Mackail, commander of Sir Derek's garrison. Within half an hour Alistair was within the Castle being wined and dined.

Later that evening Alistair delivered the King's orders to Derek and Duff. The latter stared at Alistair's hand and asked quietly, 'Do you have proof that you act for the King? Alistair was taken aback. He was unprepared for such an unenthusiastic reaction.

5

Castle Clarke—A Few Weeks Later

Half an hour of sustained bombardment was sufficient. Sir Derek appeared on the slightly damaged battlements waving a flag signifying surrender. He preferred to relinquish his guests than have his fortification destroyed. The *Providence* lowered its several boats, and Luke's men were rowed ashore. David and his small group rejoined the Scottish troops camped on the landward side of the castle. The *Providence* moved further up the loch and unloaded horses, supplies and equipment, including several cannon.

David and Luke approached the main gate and formally demanded the surrender of the castle. The gates were opened. A well built, although rather short, middle-aged man greeted them. He was dressed as any Lowland gentleman, but covered by a large woollen cape by now well spotted with flakes of snow. The defenders were disarmed, and placed in a spacious lower chamber. Some of David's troops took up a defensive position on the battlements, while others repaired the minor damage inflicted by the *Providence*.

Sir Derek led the victorious officers through the great hall of the castle and then into a small antechamber. Luke and David sat behind a small table and motioned Sir Derek to sit opposite them on a low bench. Luke assessed his adversary. Although Derek's hair was grey, and hung untidily over both shoulders, his large and equally untidy beard was still a ginger red. David was quietly pleased. He recognised Sir Derek's as the cultivated voice he had heard in the woodcutter's hut. He would not tell Luke.

Luke began the examination. 'Sir Derek, you have caused both the English military and the Scottish government considerable trouble. Whether you are executed by the military or hanged by the civil government the charge is the same—treason against duly appointed authority, and a series of other charges including murder and kidnapping. Have you any defence?'

'Of course! I have committed none of treason, murder or kidnapping. Several persons were left with me, until those who sent them here wanted their release. I simply provided accommodation for people needing protection.'

'Who brought the prisoners to you?' asked David.

'They are not prisoners. And I do not know who brought them,' Derek replied.

'Sir Derek this is a hanging matter. Co-operate or die!' threatened Luke.

'I can't tell you what I don't know. Several months ago a gentleman approached me claiming his master needed a safe refuge for persons whose lives were in danger. He would inform me when it was safe for these people to be released. He gave me an immense amount of silver to cover my costs.'

'Just how many such people are you protecting,' asked Luke emphasising the last word.

'Seven.'

'Jehu! We knew you have three, but seven!' exclaimed a surprised David.

'You must know some of these people?' asked a frustrated Luke.

'I know who each of the seven claim they are, but I have no supporting evidence of their identity. The only people I can vouch for in the other room are Captain Duff Mackail the commander of my garrison, and Lady Fenella Clarke, my wife.'

'Did you send people to Oban to murder us?' asked Luke.

'How could I? I have been unable to get messages or men out of this castle for some weeks.'

'So you have the same people here as you had, say two months ago?' continued Luke.

'As far as I know. Captain Mackail can confirm it.'

'Please ask Captain Mackail to come in,' requested David.

Duff Mackail entered with a swagger. He sat with arms crossed, silently thumbing his nose at his interrogators. He was short and unevenly proportioned. He had strong long arms, but short legs. His cavalry boots had been cut down to cater for this abnormality. He had nondescript brown hair grown excessively long, and he had no facial hair. His mouth was incredibly small, and when he smiled his face radiated menace. Luke gave an involuntary shudder. Duff's clothing was of dull olive green and brown wool. He was dressed for the climate not to impress the status conscious. His examination added nothing new to the enquiry. He led a group of clansmen he recruited at the request of Sir Derek to defend the castle. He knew all his men personally as they came from the same glen in which his father had been laird before the Campbell's had moved them out. Nobody had left the castle in recent weeks. Luke did not believe him.

David was elated. Duff was the voice with the slight Highland accent that he had heard at the woodcutter's hut. Again he would not pass this information on to Luke. His contact with the King, and his episode in the woods must remain a secret from the English. They might not understand. He had to strengthen his relationship with Luke if he were to successfully carry out his current mission.

Mackail used the opportunity to seek the release of his men and horses. Luke was for retaining the men until the situation clarified, but David supported Mackail. He argued that it would increase security in the castle, and be less of a burden on supplies. He reminded Luke that a group at the Castle had been ordered by the Black Thistle to kill him. The murderer was most likely within Mackail's band. David added that their combined forces with adequate supplies and heavy cannon could easily defend the castle.

If released, Captain Mackail could impress on others the futility of any attack on Castle Clarke without massive naval artillery. Luke eventually agreed. Next morning Mackail's men with their dirks and swords restored marched out of the castle to the tune of a sole piper, and mounting their piebald ponies rode off into the prevailing mist.

Since the capture of the castle the household and prisoners had been herded into the great hall during the day and allowed to retire to their individual chamber at night, each accompanied by two of Luke's troopers who stood guard outside their door. Over breakfast, after Mackail's departure, the officers reviewed the situation and reorganised their approach. Luke and the English would interrogate the inhabitants and be responsible for their safety; David and the Scots would be responsible for the defence of the castle. After the meal David began an inspection of the castle, and sent a patrol to shadow Mackail, and ensure that there were no enemies approaching the castle from the landward side.

Luke and Harry perused the list given to them by Sir Derek. It contained the names of three women, Elspeth Stewart, Janet Hudson and Morag Ritchie whom the chivalrous Luke immediately assumed were innocent victims; and four men – Duncan Caddell, Mungo Macdonald, Aiden Mackelvie and Malcolm Petrie–who were probably the villains. Harry was quick to add the names of Sir Derek, his wife Lady Fenella Clarke and his commander Duff Mackail to their list. Luke agreed and wondered, 'Do we have three gaolers and seven prisoners or a much more complicated configuration?'

After a minute or so of silent contemplation as both men considered the implications of this possibility Luke turned to an orderly, 'Bring Lady Elspeth Stewart to us!'

A tall thin woman in her mid twenties entered the room. Her headdress indicated she was married, and the quality of her clothing that she was a gentlewoman of the very highest status. 'My lady, we are trying to understand why you and the others are here. Tell us about yourself, and how you came to be abducted?'

'I am married into a cadet branch of the royal family.'

'Does that mean your husband is a strong supporter of his distant relative, the so-called King Charles II?' interrupted Harry.

'Please forgive my deputy. I am only interested in why you were abducted, not your political views,' Luke purred defensively.

'Colonel do not lie! Your officer is at least honest. Everything that happens in Scotland since the death of the old King revolves around how a family relates to the new King, the old Kirk and the English invaders. My husband fought for the old King and has no time for the Kirk dominated government that surrounds the new King. However my father, the Earl of Barr, dominates that very government. You can take your pick Colonel. Am I with my husband or my father? I was abducted by several masked men who kept saying that no harm would come to me, and that it was in my interests to be protected over the coming months, especially given my pregnancy.'

'Do you know why you were abducted?' questioned Harry.

'Of course, to put pressure on either my husband or my father, and to control the progress of my pregnancy.'

'So who kidnapped you? Your husband, your father or some third party?'

'I don't know. On a far more important point Colonel, all in this castle is not what it appears to be. Not all of my companions are prisoners. Some are planted to elicit information from the real victims such as myself. I do not trust any of the men. All pretend close friendship, while they question me relentlessly regarding my husband and father.'

Luke, who could hardly control his growing attraction to this confident and intelligent woman asked, 'Can you be sure that the interest of these males was not in you as beautiful woman, completely free of any political or conspiratorial connotations?'

'Colonel, all men are not like you. Your soldier's eye may see me as a woman on whom to exercise your charm, but to my Scottish male companions and relatives I am simply a pawn in their political power game.'

Lady Elspeth was indeed a beautiful woman whose long gingerish hair was not completely concealed by her curch. She had a long face whose large pale blue eyes were very close together with a long thin nose that ended very close to a small mouth with a thick lower lip and a very thin upper lip. Luke regained his composure and asked,

'As you do not believe all of the men are prisoners what can you tell us about them?'

'Not much. I avoid any one-to-one conversation. I make sure that Lady Fenella or young Janet Hudson is always present. However over dinner the men feel the need to boast about their achievements and position, and when challenged by each other are often indiscreet.'

'And what have they revealed?'

'Why should I tell you?' replied the spirited gentlewoman.

'Major Burns and myself were ordered by the English and Scots governments to rescue you, and return you to your family. They were informed that what was happening here was relevant to a major plot against specific members of both English and Scottish administrations. So the second part of our mission is to uncover as much as we can about this impending disaster, which in essence involves discovering the members and plans of an organization called the Black Thistle,' explained Luke.

'All that I can say is that the information revealed by the men in this castle is contradictory and confusing.'

'In what way?'

'Sir Derek is the exception. His interest is simply to increase his influence within the area—and he doesn't seem to care who controls Scotland in the future, the King, the Kirk or the English. That is why he readily surrendered the castle. He did not want it damaged. The *Providence's* firepower must have terrified him. He was also delighted that you allowed his soldiers to leave the castle unharmed.'

'And Sir Malcolm?'

'Sir Malcolm Petrie claims to be an elderly scholar who has very recently been in France and Holland. The names he mentions in passing are known Royalists, barred from participation in Scottish affairs. He dislikes me, but I do not know why.'

'So Petrie may be a Royalist agent in contact with powerful Royalist refugees in Europe?' posed Luke.

'Possibly. Andrew Mackelvie is a local laird representing Campbell expansion into areas to the north. He spends his time bemoaning the division between his clan chief, the Marquis of Argyle, and Argyle's son, Lord Lorne. He was careful not to express which of the family he supported.'

'And what about the clergy?'

'Duncan Caddell is a horrible little toad. He is a minister of the Kirk and expresses the most extreme sentiments, and deplores the decline in public morality. Mungo Macdonald also claims to be a minister of the Kirk but by Caddell's standards he is a loose living sinner. Macdonald is an impostor. I am sure I have seen him before.'

6

Janet Hudson was very short. She had long flowing black hair that reached her waist and dark brown eyes sunk deeply into her head which almost circular in shape. She had a snub nose, large mouth and very little chin. She wore a deep green bodice and a similarly coloured short overskirt that revealed a cream underskirt patterned with crisscrossing lines of red and green. On her breast she pinned a large silver broach studded with large emeralds and smaller diamonds. Her cuffs and collar were of fine cream lace. This was a wealthy young lady of about eighteen years whose elaborate dressing was in sharp contrast to her narrow and moralistic outlook on life. Her red cheeks reflected her nervousness at being alone with two English officers–with two men.

'Miss Janet why were you kidnapped?' asked Luke.

'I have no idea. My brother was about to announce my betrothal to a relative of the Marquis of Argyle. My abduction stopped this. To be honest after talking to Lady Fenella I am not sure that I was abducted. Maybe my brother sent me here for my own protection.'

'To protect you from what? Do you have strong political or religious opinions?'

'I take no interest in politics, but I am a devoted to the Kirk.'

'And a supporter of the so-called King of Scotland?'

'Sir, if the King takes the Covenant and defends the true Kirk, then with all Scots I will be his fervent follower. God created monarchy in His image. In killing the old King, you English broke God's word, and committed sacrilege as well as murder.'

'Were you kidnapped to stop your betrothal to a Campbell?' asked Harry, a little frustrated by Luke's gentle questioning.

'There would be many in my family who would not like to see such a marriage, but my brother and senior uncle favoured it.'

'What about yourself? Do you want to marry your intended?'

'A strange question sir. I have never met him. I do whatever my brother wishes. God has placed me on this earth to do His will, as outlined by my brother.'

'Have you discovered anything about the men in this castle that might explain why you are here?' asked Luke.

'Lady Elspeth hates them all, but I don't. Mr Caddell is one of God's ministers, and is a great comfort. Mr Macdonald is also one of God's ministers but Mr Caddell said he is a scandalous man and must be avoided. But he has been kind to me. Master Mackelvie is related to my betrothed, and we have discussed matters regarding my future husband's family and friends.'

'Does he support your intended marriage?' enquired Luke.

'He said there was opposition from many Campbells, but strong support from those closest to my intended. Of the others Sir Malcolm seems a nice old man who talks to me about the merits of the Geneva Bible.'

'Thank you Miss Janet, you may leave,' announced a suddenly bored Luke.

'Don't you want to hear my views on Sir Alistair, and my female companions?' asked Janet who was now quite enjoying her interlude with the English officers.

Luke was stunned. Sir Alistair? His name did not appear on Sir Derek's list. He could hardly conceal his interest, 'Tell me about Sir Alistair! What was his family name?'

'He never revealed it. He was quite mysterious. He was the nicest of them all. He was also the most powerful. The others showed him considerable deference, although he did not seem to relate well to any of the men. I assumed he represented the Government. He promised me my freedom.'

'Have you seen him today?'

'I have not seen him since your arrival. Given his obvious importance, I thought you had arrested him.'

As soon as Janet departed Luke ordered another search of the castle and confronted Sir Derek over his omission. He was unfazed. He had only provided a list of those currently in the castle. He assumed that Sir Alistair has escaped during the attack. Luke returned to his enquiries and called in the young child, Morag Ritchie. A striking blonde whom Luke knew to be Lady Fenella Clarke accompanied the girl explaining that, 'Young Morag is scared of the horrible English. I hope you don't mind me coming with her.'

'Not at all Lady Fenella. After I finish with Morag I would like to ask you a few questions.' Luke turned towards the girl of ten or eleven years. She was a plain child with an oval freckled face and a couple of protruding front teeth–in appearance a most unappealing child.

'Don't be frightened Morag. Tell me about your parents.'

'They died when I was very young. I live with people who I call uncle and aunt, but they are not my real relatives. They are my guardians, but my property–which I inherit when I become older– is administered by one of the great nobles of the land.'

'Do you know why you were abducted and brought here?'

'I am not sure that I was. My uncle told me that I was to go on a trip ordered by the Earl of Barr, the guardian of my lands. Uncle's men escorted me to the nearest town from where another group of horsemen brought me here.'

'Thank you child. You may leave us while I talk to Lady Fenella.'

Luke turned to the mistress of the household, 'There is not much mystery surrounding the abduction of the child, only the identity of the kidnappers remains to be solved. Do you know anything about Morag's background?'

'The death of her parents was a major crime several years ago. A gang broke into the family's house and murdered her parents in front of her. She has no memory of the atrocity.'

Lady Fenella Clarke had discarded her headdress allowing her long blonde hair to trail over her shoulders. She wore a pale blue bodice with a matching long overskirt, both embossed with much silver thread. The expanses of fine lace around collar and cuffs suggested that Sir Derek spent much of his ill-gotten gains on his wife. On her breast she displayed an intricate silver broach with a large blue sapphire that competed with her sparkling blue eyes. She was in her mid twenties, about half the age of her husband. Her lips were wide, and when she smiled her eyes and mouth acted together to exude an overt sexuality. She was immediately responsive to Luke's unconcealed interest. 'Colonel, don't look at me like that! Elspeth said you were a womaniser.'

'Forgive me Lady Fenella but you are a beautiful woman.' There was silence as both parties struggled to contain their growing mutual interest. Luke lowered the temperature, 'Now, what can you tell me about your female companions?'

'Chalk and cheese. Lady Elspeth is a woman of the world. She is a match for most men. Someone will pay dearly for her abduction. Miss Janet is naïve and trusting. She believes whatever the clergy or her brother tell her. She is a real innocent.'

'What about the men?'

'A boring bunch. Until your arrival Duff and Alistair were the only real men in the castle.'

It was Luke's turn to feel discomforted by Fenella's hardly concealed sensuality which he struggled to ignore and, almost in desperation, asked, 'Tell me about the mysterious Sir Alistair. Where is he?'

'I hope he has escaped, but I have a bad feeling. He is a powerful figure. The men showed him great respect. He disconcerted my husband. Initially I thought he might be behind the kidnapping. Then he questioned everybody relentlessly—much more incisive than yourself Colonel. And he told Elspeth, Janet and young Morag not to fret, because they would soon have their freedom. He was particularly sensitive to Elspeth's needs.'

'Perhaps he is behind all this. It takes a powerful man to have both the English and Scottish governments send an expedition to solve the problem. After all we are technically at war with each other, and my General is just waiting his time to attack the Scottish army at Stirling. Maybe Alistair has disappeared with all the information he needed from the prisoners,' concluded Luke.

'I hope so, but I have a bad feeling, and I have evidence to support my fears,' moaned Fenella.

'What evidence?'

'The night before you attacked I heard my maidservant screaming. I ran upstairs and found her crouching against the wall. She pointed to the stairs that led to the upper levels. There, on the steps, was a green apparition—a headless woman, carrying her head under her arm. I have seen her before. According to my husband she has haunted the castle for centuries.'

'How is it relevant to Sir Alistair?'

'The ghost appears when a death is imminent. I have not seen Alistair since I saw the ghost. His may be the death foreshadowed by its appearance. Alistair might have been murdered. Although everyone showed him great respect the men were humiliated and embarrassed by his public interrogation of them. Mackelvie and Caddell badmouthed him constantly.'

'What's your opinion of those men?'

'Caddell is a dangerous fanatic. He would destroy the culture of the Highlands and enforce a narrow moral society in which the clergy would have power to enforce morality, and punish breaches of it. His society would be worse than that created by an English military government. I hate the way he manipulates young Janet.'

'And Aiden Mackelvie?'

'A mean spirited neighbour of ours who as a Campbell thought the future was his. Now he is full of self pity as his clan chief Argyle, and the chief's son Lord Lorne are sworn enemies. Lorne supports the King unconditionally, whatever his father does. Mackelvie does not know which way to jump.'

'Mungo Macdonald and Malcolm Petrie?'

'Macdonald is an impostor. He claims to be a minister of the Kirk but he avoids talking theology and moral reformation with Caddell, despite the latter trying to provoke him. His knowledge of the Scriptures seems especially sparse. He has been caught out on several occasions.'

'Who is he then?'

'A Catholic priest? The Highlands are full of Papists. He is certainly someone's agent with his own agenda. Duff distrusts him. And Sir Malcolm Petrie is not the pedantic old scholar that he tries to project. He is a well educated, politically astute diplomat, who has been on missions across Europe for a decade or more.'

'Is he the Black Thistle?'

Fenella hesitated and avoided the question. 'All I know about the Black Thistle is that when I asked Derek who was paying him to guard our guests, he said it was the Black Thistle.'

Fenella invited Luke and other senior officers to eat with the household and its guests. That evening Luke and David sat at a long oak table, with the Clarkes occupying their traditional position at either end. After Cadell indulged himself in a mini- sermon disguised as grace, Sir Derek's servants brought to the table a vast array of steaming meats, boiled mutton, roast lamb and various baked pies of venison, chicken and lamb. Sir Derek did not stint on expenditure when it came to presents for his wife, or food for his guests. Luke, having previously gained a liking for Irish whiskey, was introduced to its Scots cousin—a drink which quickly became the antidote to the freezing weather.

David decided to disconcert the diners. 'Where is Sir Alistair?' he asked.

Surprisingly Duncan Caddell responded eagerly. 'The man was an embarrassment. He interrogated us on the most personal of matters. A highly dislikeable individual. God respects privacy, so should man. He obtained all he could from our little group and departed to report to his masters.'

'And how could he do that given the blockade?' asked Luke.

'The blockade never worked. Sir Derek has contacted the outside world whenever he wished,' replied Caddell, with a withering glare at his host.

'Quite true,' responded Derek, who seemed completely unfazed by his earlier lies. 'Captain Mackail and I communicated with our friends outside the castle. Consequently Alistair either left before your attack, or marched out disguised as one of Mackail's men.' Luke was infuriated at the smile of quite satisfaction that Derek made no attempt to conceal.

He was stunned by Fenella's intervention. 'Alistair probably escaped through one of the several secret tunnels, that according to tradition, honeycomb this edifice.' She enjoyed the effect of her contribution on the gathering.

'You didn't mention secret exits Sir Derek', remarked an angry Luke.

Fenella again intervened, 'Derek knows nothing about them. We are recent occupants of this castle. It once belonged to the Cameron clan, and the then young Duff Mackail stayed here as a boy. When you attacked he asked me did I want to leave, and explained he could take me to safety through a secret passage. Duff may have made Alistair a similar offer.'

7

'**B**ut your ladyship only today expressed fears for Sir Alistair's safety?' queried Luke.

'True. The apparition worries me. We are all here, except for Sir Alistair. My heart tells me something horrible has overtaken Sir Alistair, but my head suggests there is a logical explanation–his escape through one of the secret tunnels.'

David asked, 'Does your ladyship know where these secret tunnels are?'

They accepted Fenella denial, and then Luke asked, 'Gentlemen, if Sir Alistair is dead can any of you suggest a reason?'

Sir Malcolm unexpectedly took offence. 'If Alistair is dead then your implication is that one of us, or Captain Mackail is responsible.'

'And are you?' Luke, with emphasis, slowly enunciated. 'Alistair annoyed and perhaps terrified some of you with his incisive questions. Others he humiliated. You had good reasons to kill him.'

'And how would we do that? squeaked Aiden Mackelvie. 'We were all disarmed, and Captain Mackail had one of his men with each of us at all times. Sir Alistair was a big man, and strangely was not disarmed. As you can see, all of us are small or aged–no match for a man in his prime.'

'So the consensus is that Sir Alistair has left the castle, and not fallen victim to foul play?' summarised Luke.

'Not at all!' announced Mungo Macdonald. 'I overheard Caddell and Mackelvie plotting their revenge on Sir Alistair. If he is dead, ask them!'

Mackelvie's face expressed pure venom as he turned on Macdonald, 'You lying churl. You are no man of the Kirk. If anybody had anything to hide from Sir Alistair it was you. Who are you?'

'I am a Macdonald on a mission for my clan chief when I was abducted. Remember, you Campbell-loving traitor, that there are large sections of the Kirk that reject the narrow mindedness of your toady Caddell and his ilk. I am surprised that two such Campbell vipers were abducted. You have nothing in common with the rest of us true Scots.'

'So what do we have here then?' asked Luke. 'Sir Alistair uncovers a Campbell plot on the one hand, or a Macdonald one on the other, and is murdered before he can report to his master, whosoever that might be.'

Petrie changed the topic. 'Colonel, your mission was to free us, but you have made no moves to do so. Why is that?'

Luke replied, 'Both the English and Scottish authorities received intelligence that a plot to affect the government of both states by eliminating their leaders was being hatched. More specifically the inhabitants of this castle were involved. Until we uncover the extent of that involvement you remain here. I don't know why most of you were abducted. Some of you may not have been kidnapped at all, but are agents of the abductors who placed you here to spy on the others. Young Morag was kidnapped because she is heir to vital property in a rapidly escalating factional conflict. Even here I don't know whether she was taken to protect her from her enemies, or to assist an usurper replace her.'

The women who, apart from Fenella, had been quiet throughout the dinner withdrew, as did Caddell. He would read the scriptures to Janet. The remaining men gathered around the fire and Sir Derek dispensed large draughts of excellent whisky. He then candidly admitted, 'My role in all this has been commercial. As a business undertaking I agreed to hold people in safe custody until those who paid me determined otherwise. That contract has been nullified by the military intervention of the Scottish and English governments. Therefore I advise you to co-operate with the officers so that they will agree to your release. I have not been paid to provide for you any longer.'

Luke indicated that he would continue his interviews in the morning and hoped that an evening of heavy drinking might loosen tongues. Luke circled the room topping up each man's drinking mug. Mungo Macdonald intrigued him. Although he was dressed in clerical black his large white lace collar and immense lace cuffs seemed out of place. He was bald but had a carefully cultivated curly black beard which he was forever stroking. He was in his late forties and well built with strong looking arms and large hands. Surprisingly he openly carried a dirk despite claims by others that they had been disarmed. He was a heavy drinker, but still resisted all attempts by Luke to elicit relevant information.

Mackelvie was distinguished above all by a high-pitched voice. Those who only heard the voice would have sworn that it belonged to a woman. He was a pompous man of medium build, with a rounded face and light brown hair worn short, and blue eyes set well apart from each other. They were separated by a large flattened nose. Luke surmised that it may have been broken in the past, and had not correctly healed. His doublet and breeches were a matching dark grey brightened by a large gold medallion that he wore on a very thick gold chain. There was the minimum of lace at collar and cuff. Unlike Macdonald he drank little, and Luke caught him pouring his mug into Macdonald's cup when the cleric's attention was diverted. These two men had secrets that Luke was determined to uncover.

Next morning Luke's orderly escorted Sir Malcolm to the interview chamber. Sir Malcolm was an elderly, if not ancient gentleman. His grey hair was worn long and his long wide beard, a mixture of silver grey and white, was meticulously trimmed. He dressed totally in black but what distinguished him from the clerical community was the excess of jewellery that adorned his person. Around his neck he wore a gold chain and his fingers were covered in gold rings containing every conceivable precious stone. Age affected his ability to concentrate, as from time to time he stared vacantly into space as if in another world.

Luke was gentler than he had been the previous evening, 'Can you enlighten me on your abduction?'

'It is not related to high politics, despite the dilemma all Scots currently face–if Kirk and King fall out which way do we jump? As I have not expressed any views on this matter, I cannot see how my abduction is related to it.'

'You are not a member of any specific faction?'

'I am with a majority of Scots who hope that King and Kirk can work in unison. Therefore I can only assume that people who wish to break the alliance between King and Kirk would want me out of the way. The only people who benefit from my abduction are the English.'

'Are there individuals in high places that would want you out of the way if they decide to go with the King without the Kirk, or with the Kirk without the King?'

'Kirk without the King is not an option. It is a question of degree. How much power would a Kirk-dominated regime allow the King? The King is appointed by God, and is God's image on earth.'

'Are you in any way connected to any of the other prisoners or their families?'

'Twenty years ago when I lectured at University apparently Duncan Caddell was one of my students. He claims to know me, but I have no memory of him. I am acquainted with Lady Elspeth's father for whom I conducted many diplomatic missions. Of the others I have no acquaintance, although Mungo Macdonald is vaguely familiar.'

'Have you heard of the Black Thistle?' Sir Malcolm paled, 'Yes.'

'Who leads it?'

'Somebody close to the King. Letters were sent to my former colleagues at the University warning them that their positions depended on not siding with the English. I was to advise them how to react to this threat when I was taken.'

Duncan Caddell was the next to be interrogated. Luke disliked him. He was a small, almost hunchback figure, dressed completely in black, with the smallest of lace collars, and barely a glimpse of lace cuffs. He was a thin man with a thin long face and a double chin. He wore his dark brown hair cropped very short and his deep brown eyes penetrated those with whom he established eye contact. He used his eyes to unsettle his adversaries. He stared at Luke–who stared back– and began a long diatribe against the English religious reformers. 'The Holy Writ has laid down a blueprint for life on this earth, and in preparation for the next. There is no room to question God's word. There is one God and one true Church and the Kirk is that Church. And the King is God's representative on earth. Other views are in error and their holders must be disciplined.'

'But if you are forced to choose between a King who does not accept the Kirk, and the English who tolerate a range of views you despise, which way do you go? In the next few months you will have to choose between supporting the so-called King or the English invaders, both of whom reject the authority of your Kirk.'

'God forbid that such a choice is forced upon me.'

'Come on Caddell. You are no fool. This issue is on everybody's mind since Dunbar. Will you support us, or Charles Stuart?' Luke persisted

'I will not answer such a question.'

'Why were you abducted?' asked Luke, attempting to loosen up the tight-lipped cleric.

'The sinners of this world would like to silence my voice for a continuing moral reformation. Clan chiefs, great nobles, hundreds of lairds, courtiers, and soldiers are all doing the Devil's work–and enjoying it. My murder would serve their ends, but my abduction seems meaningless, unless it is a prelude to my death.'

Caddell suddenly paled as he realised the implications of what he had just said. He contemplated his own mortality as Luke prosaically continued his questioning. 'Have you had any links with any of the other prisoners or their families?'

'Sir Malcolm Petrie taught me theology at university twenty or more years ago before he moved into law. I have met Aiden Mackelvie during my ministry in this part of Scotland, and I am sure I have seen Sir Alistair, but I cannot remember in what context, but he claims he has never met me.'

'You have no links with any of the three women?'

'No, two of them are great sinners. Lady Elspeth is a temptress, and used by Satan to achieve his ends. She has moved away from the true teachings of her father. Beware of her colonel. Young Morag I suspect is an illegitimate child of some powerful sinner. On the other hand Miss Janet Hudson is a daughter of the Kirk for whom I have become a mentor.'

'And has this mentoring led to any revelations from the young lady?'

'No, she is so naïve politically that I cannot see why she was kidnapped, unless it was to prevent her inappropriate betrothal.'

'Have you heard of the Black Thistle?'

'Yes, an extreme Royalist organization.'

'How do you know the name?'

'In visiting the homes of many nobles and lairds in recent months I have heard them mentioning the name as a rallying call against surrender to the English.'

Before Luke could probe further an orderly flung open the door and blurted out, 'Colonel, the young girl, Morag, is dead.'

8

Luke ran to Morag's chamber where Fenella was cradling the young girl in her arms. Luke could see a loosened ribbon around the girl's neck, and red marks where it had bitten into her throat. Luke placed the shining blade of his dagger across the girl's mouth and shouted, 'She's alive. Is there any one in the Castle with medical knowledge?'

Fenella indicated that one of her servants was an expert with herbs and potions. David Burns, who had just arrived in the chamber, sent for his deputy, Dugall Sinclair, who was trained at the University of Edinburgh and had considerable experience as a battlefield surgeon. Janet Hudson offered to sit beside the girl.

Luke turned on his troopers. 'How did this happen? You left your post outside the door of this chamber?'

'No. We have just relieved the night guards. Speak to them,' an irate soldier replied.

Luke did. The guards denied having moved from the door. They had escorted the girl to her chamber after dinner, and a servant of the household brought her a drink later in the evening. A trooper was present while the servant was in the room. He then resumed duty outside the door with his compatriot, until relieved an hour earlier.

Lady Fenella shed light on the situation. 'Young Morag did not appear for breakfast. I thought she might be ill. Your trooper led me into her chamber, and I found Morag lying on the floor. I thought she was dead, and sent the soldier to fetch you.'

Luke examined the girl and the room. He announced largely for his own consumption, 'What happened is clear. Someone tried to strangle the girl. She fainted, and the would-be murderer thought he had achieved his purpose. He left the unconscious girl for dead.' Luke was silent for a while and then exclaimed with frustration, 'But how did he get past my troopers?'

Fenella drew Luke aside, 'When Duff Mackail suggested I leave the castle to avoid the unspeakable abominations of an English occupation he showed me the entrance to one of the secret tunnels. It begins behind the hearth in Morag's chamber.'

'So you lied to me about your ignorance of these secret tunnels. The situation does not look too promising for you. Whoever tried to kill Morag knew of this tunnel. From what you tell me it could only be Duff Mackail, Sir Alistair or yourself. Both Duff and Alistair are not here.'

'Who knows which other members of the household, or our guests have knowledge of this tunnel,' Fenella quickly replied.

Luke and Harry set off to explore the tunnel. Armed with a large taper they entered Morag's room. She was still being watched over by Janet Hudson who nodded at the soldiers as they manoeuvred the large blocks at the side of the hearth. One of the rings imbedded into the wall proved to be the handle to effect the opening. Once the hidden door had opened Luke and Harry found they were not in a tunnel but a large room, part of which received natural light through windows that looked out onto the loch. The room contained a large bed that had been recently used. 'Which of our ladies escapes here with her lover? And who might he be?' sniggered Harry.

Luke muttered with a tinge of jealousy, 'Lady Fenella and Captain Mackail.' The men moved to the hearth of the hidden bedroom and noticing a similar ring to that in Morag's, they opened yet another hidden door. They were now in a tunnel. After ten minutes the men saw a glint of light in the distance. Once they reached the source of the light they discovered a number of small steps leading to a partially demolished trap door. Moving the door aside a large amount of snow fell in on Luke. Emerging into the muted daylight they found themselves within a ruined cottage in the middle of a heavy snowstorm.

Luke did not know where they were in relation to the castle as visibility was reduced to a few yards. He moved a few stones into the shape of a cross. He would ask David to locate the hut with a surface patrol. The cold was beginning to bite and the two men quickly headed back up the tunnel. Suddenly there was a gust of wind that extinguished their tapers. They were in complete darkness. 'Thank goodness this is a straight tunnel. We can feel our way home,' Harry announced happily.

'Just to make sure one of us does not fall over or gets confused take one end of my bandolier,' ordered Luke.

The two English soldiers proceeded slowly as both remembered that the floor was uneven and there were several major hollows. They suddenly stopped, alerted by the same phenomena. In the distance, but coming closer they saw a lighted taper. Luke whispered to Harry, 'Press yourself hard against the side of the tunnel. It too is uneven. Try to conceal our presence until the light carrier is almost upon us.'

The figure gradually approached and at the last minute sensed the presence of the English soldiers. He drew his sword. Before he could speak Luke announced, 'I wouldn't do that sir. There are two swords and two daggers ready to confront you. Please continue on your way. Our taper was extinguished by the gust of wind caused by your entry into the tunnel.'

Luke and Harry allowed the man to lead them back to the castle. On entering the hidden room the Englishmen were astonished to find a scantily clad woman already occupied the bed. It was Lady Elspeth. Harry whispered to Luke, 'We got the wrong woman Colonel. We should have had a wager on it.'

Elspeth was clearly pleased to see the stranger, and was not particularly affronted by his two companions. Luke saw the political advantages of the situation. 'I can easily forget what I see if the two of you help me separate the victims from the conspirators, and identify the members of the Black Thistle.'

Lady Elspeth drew a cape around her shoulders assisted in the endeavour by the clearly besotted male. She smiled sensuously at Luke and said, 'I did not expect a professional soldier of your experience to be such a prude. I know from the way you leer at me that you are not. What you see is but domestic bliss. Colonel Tremayne of the English army meet Sir Alistair Stewart, my husband–the elusive Sir Alistair.'

Sir Alistair was an impressive figure. He was tall and solidly built. He had long straw-coloured hair that would make him stand out in any company, and an olive complexion. His blue grey eyes twinkled. He exuded a cordiality to which Luke wanted to respond. Although his sword had a bejewelled hilt, and he was dressed as a courtier—with burgundy doublet with light blue silk slashes and matching breeches—he had the confidence of an experienced soldier. Luke drew in a large breath, 'Why the subterfuge?'

Sir Alistair replied, 'When Elspeth was abducted her father's men traced her to this castle. Elspeth's father, having discovered where she was held, seemed content to leave her here. I was not. I used my position with the King to be sent here.'

'So why did you not free her and the two of you leave the castle?' asked Harry.

'As soon as I ascertained that Elspeth was in no immediate danger I tried to uncover for the King what exactly was going on here. The guests are a collection of strange characters with diverse political and religious views.'

'Can you separate the victims from the conspirators?' asked Luke.

'Not completely. I heard Macdonald comment to Duff Mackail that Scotland might have to exist without King, nobles and the English generals. It suggested to me that someone aimed to govern Scotland by eliminating these possible rivals. I sent a man to warn the Scottish government. He was attacked soon after leaving here but lived long enough to give his message to an English patrol. Your government informed the Scots, and your joint expedition is the result.'

'Why send the warning? As a Royalist the death of Scottish officials opposed to the King, and of English generals intent on destroying Scottish independence is in your interests,' asked Luke.

'The King is no fool. He is about to be accepted as the real ruler of Scotland and does not wish to tarnish his reputation with illegal or immoral acts. There are some maverick loyalists whose activities he needs to disown.'

Luke asked, 'If Black Thistle is planning a simultaneous assassination of Scottish and English leaders what's the point of having five people locked up in this castle miles away from the centres of government?'

'That is the real mystery Colonel. Nobody here can be a frontline conspirator. They are being held to persuade others to do Black Thistle's bidding.'

'So by keeping the guests here we are playing into Black Thistle's hands?'

'Yes, you would do better to release everybody, and then monitor their subsequent behaviour.'

'Maybe, but it could be a risk. Someone like Caddell or Mackelvie could indeed be a potential assassin,' concluded Luke.

Luke met his officers at breakfast. Harry was frustrated. 'Luke, we have an impossible task. We know nothing about the people in this castle, and you cannot believe what they say about themselves, and each other.'

'What are you suggesting?'

'Conduct a proper investigation, and find out all we can about them.'

'And how do we do that? We are in a foreign country and one that we are about to conquer. Some of our potential sources speak a language we do not understand. Are such people likely to tell us the truth?'

'You are a pessimist. Most of the people we need to question are Scots speaking, not Gaelic Highlanders. Most come from the eastern Lowlands which is already under English military occupation. The Scots are divided over family feuds, religious principles and attitudes to the monarchy that sufficient numbers will tell us all we need to know.'

After two days of examining Mackelvie and Macdonald, and re-examining the others Luke had unearthed no more useful information. Next day Luke, Andrew and David's deputy, Captain Dugall Sinclair left for the Lowlands. They eventually arrived at the manor house of Morag Ritchie's guardians. A servant led them into a small chamber off the main hall where a very elderly man and his equally elderly wife rose to welcome them. The dress of the couple indicated that they were extremely wealthy, as did the furnishings of the room. The man spoke, 'Welcome gentlemen! What news of our ward young Morag? The Earl of Barr who administers her estates sent us a message to say that she had been located, and he was taking steps to free her.'

Luke replied, 'A combined English-Scottish force captured the Castle in which Morag and several other prisoners were being held. We are unable to release her until the role of these prisoners is fully understood, and the reasons for their captivity clear. We certainly do not want to release her into even greater danger.'

Dugall added, 'That is why we are here. To tell you that Morag is safe and well, and to find out more about her abduction.'

'What do you know so far?' asked the old man.

'Morag is heir to an immense fortune. Her parents were murdered when she was very young. She has lived with you ever since,' answered Luke

'Morag is heiress to a vast fortune of which this estate is part. Until she becomes of age or marries it is administered on her behalf by the Earl of Barr. The Earl appointed us her guardians. Even though she is still very young the Earl was taking steps to arrange a betrothal with one of his political allies. Morag's real family have never been close to the Earl of Barr and some would stop at nothing to prevent her premature betrothal. They do not want the property to leave the family and finish up in the hands of young Morag's future husband, a husband determined by the Earl of Barr.'

'How would kidnapping achieve that end?' enquired Dugall.

'The Earl would be unable to effect his end without her physical presence and given the political turmoil her abductors only have to wait a few weeks and the Earl will be driven from power. If you had not rescued her I am sure she would have come to no harm. Her kidnappers would simply have replaced the Earl in the supervision of her property, and then married her to one of their supporters.'

'Morag told us that your own men carried her away,' said Andrew, hoping to catch the old man off guard.

9

The old man eyed Andrew warily. He was not fazed. 'True. My steward was told that the Earl of Barr had ordered Morag taken to the nearest town where they were to hand her over to the Earl's own men. The Earl later denied he gave any such order and declared that the men were definitely not his.'

'Despite the Earl's denial he could have ordered her kidnapping to protect his own interests,' Luke commented. 'You said her real family were opposed to Barr.'

'Yes, her late father was a close friend of the Earl, but her surviving relatives are fanatical supporters of the King. Morag's properties would add greatly to his cause.'

The old man slipped into a silent reverie. His wife appeared anxious and whispered to him, 'Tell them everything, my husband. Young Morag is still in danger.' He indicated that she could explain further.

The old lady with a soft, low-pitched voice took up the tale.

'When Morag was little she often crept into her parents' bed. One terrible evening the house was invaded, and Morag's parents were killed where they lay. The murderers spared the girl probably because she was too young to be a threat.'

The old man commented, 'I thought the murderers were family who would benefit by Morag's minority. The Earl of Barr quickly put an end to any such gain by appointing himself as administrator of her properties, and ourselves as guardians of her person, much to the chagrin of her own family.'

The old woman's anxiety did not subside. She spoke slowly, emphasizing each word, followed by a long delay. 'Property is not the issue.' After a minute or so she elaborated, 'Just before she was abducted Morag began remembering things from her past. The trauma of the murders had suppressed details from the period for almost a decade. She now remembers that when one of the masked men removed his gauntlet to check the bodies of his victims, he could turn his hand back upon his wrist.'

'God's Blood!' exclaimed Luke. 'The man who took part in the murder has an abnormality that can be traced. He would not want that known. Let's hope he does not become aware of Morag's returning memory.'

'And let's hope that Morag has not told anybody at Castle Clarke of this development,' added Andrew.

After completing the midday meal with Morag's adoptive parents the party moved on to an area where Duncan Caddell had been a minister. They stopped at a rather large church. A tall gaunt man dressed in the sober garb of the clergy nervously greeted them.

'I am a man of God, a man of peace. Why are three armed soldiers in my churchyard?'

'Fear not Reverend! We bring you good news, and seek further information. Your fellow pastor, Duncan Caddell, has been found, and is currently safe. However in order to protect him, and release him, we need to know his enemies, and why he was abducted.'

The minister gave a nervous laugh, and motioned the soldiers into the church where they took a seat amongst the pews. 'Caddell has hundreds of enemies. He is the most hated man I know. It was intense two or three years ago, but the antagonism has hardly subsided. When he disappeared most people around here hoped he had been killed, and rejoiced.'

'Why would a single cleric arouse so much dislike?' asked Luke.

'Not dislike, bitter hatred,' spluttered the minister.

Dugall answered Luke's question. 'Now I remember. Our Mr Caddell is the Caddell who led a violent and vicious witch hunt a few years ago. He is a brutal witch hunter.'

The minister took up Dugall's story. 'Due to Caddell's fanaticism a minor witch investigation got out of hand. Dozens of innocent people were accused, tortured, and several executed. Caddell represented the extreme fundamentalist party within the Kirk who dominated in this area until the recent English occupation.'

'Are the records of the witch trials undertaken by Caddell held locally?' asked Dugall.

'This is the senior church within the local synod. The interrogation was held here. I have the records. Follow me!'

The soldiers were soon perusing the records of witchcraft prosecutions, all signed by Simon Caddell. Luke turned to the minister, 'Sir was there one case that in particular infuriated the locals–a case where popular opinion was on the side of the victim?'

'The Lady of Dalmabass! The big house that you can see further along the river is the residence of the Laird of Dalmabass. The previous laird remarried late in life to a much younger woman. Within a month the old laird took ill and died. He was eighty-four. While the lands went to his eldest son, the current laird, all his other wealth, which was considerable, went to his young widow. The new laird was furious and determined to get his hands on his father's moveable possessions.'

'So he accused his step mother of witchcraft?'

'Not initially. The first charge was murder–that the second wife had poisoned her husband. Her lawyers very quickly had that charge dismissed. It was Caddell who suggested to the new laird that he charge his stepmother with causing death through witchcraft.'

'And Caddell became chief prosecutor and judge?' asked Dugall.

'Yes, and as the interrogation progressed it was obvious to rational gentlefolk that Caddell was obsessed with the Lady Dalmabass.'

'Surely a woman of her class was not subjected to torture.'

'I'm afraid in destroying Satan's instrument Caddell left nothing to chance. She had a rope tied around her head and it was then twisted tighter and tighter as it cut into her flesh. Finally Caddell had her ladyship's fingernails ripped out. If it had not been for the intervention of local gentry she would have had needles inserted in place of the nails. During the frenzy of the witch craze the common people temporarily went along with Caddell. When the situation quietened down and a few of the witnesses confessed they had been cajoled to give evidence by Caddell and the current laird, the truth finally emerged.'

'So the current laird is not popular?'

'Half of his house is in ruins. Someone set fire to it only last week.'

'What happened to the Lady of Dalmabass? Was she executed?'

'No, she was convicted and confined in the tower of this church awaiting transport to Edinburgh for confirmation of the sentence– and execution. The local guards were bribed, and she was rescued by her brother. He disappeared with her into the Highlands where she remains to this day beyond the reach of the law, and the Kirk.'

'Do you know her family name before she married the laird?'

'No, she was married in Edinburgh so the local records are silent, but she told me when I took her food during the trial that she came from Cameron lands in the Western Highlands.'

The soldiers accepted the minister's offer to spend the freezing night in his warm barn. Luke's relationship with Dugall became as icy as the weather. Luke's plan for the following day was to question relatives and servants of the Earl of Barr. Dugall objected, 'The Earl represents the government of Scotland, and is my commander-in- chief. It is disrespectful, and not politically wise to embark on such an enquiry on his personal estates.'

'Then what do you suggest?' asked a slightly perplexed Luke.

'It's a little further but Sir Alistair Stewart needs more urgent investigation. A number of lairds to the south of here are Stewart's cousins with whom he had often clashed.'

As Dugall rode on ahead Andrew muttered to Luke, 'Are our Scottish friends here to help with the investigation, or protect the Earl of Barr?'

'We will investigate Barr later–without our Scottish watchdog,' promised Luke.

The number of English patrols increased as they headed southeast. Luke sent a coded message by one of them to Cromwell expressing his concerns about his Scottish partners. Around midday Dugall indicated that the large house just off the road and surrounded by a dense woodland belonged to one of Sir Alistair's estranged cousins. As they made their way along the curving drive towards the house a number of servants gathered around them and shouted abuse. Luke soon realised that it was not directed at Andrew or himself, but at Dugall. When they reached the entrance to the house the steward of the estate barred their progress. He bluntly declared, 'The Earl of Barr's man is not welcome.' The servants now formed a solid phalanx around Dugall pressing against his person.

The steward with a wave of his hand calmed the mob and told Dugall, 'Do not fear. No harm will come to you while your comrades discuss with my master whatever business has brought them here.' Dugall reluctantly agreed to stay outside after which Luke and Andrew were ushered inside to a small chamber. They were soon joined by their host whose first thought was to provide hospitality. 'Given the hour, you have probably not eaten for some time,' he correctly surmised. He clapped his hands and food and drink were brought to his English visitors. Luke was surprised, and could not contain himself. 'Sir, I am astonished that you welcome us with generous hospitality, yet reject a soldier of your own army in a most objectionable way.'

'No one associated with the Earl of Barr ever put any interests above that of the Earl. He will soon be gone and the King will rule. At least here, with English military dominance, we can go about our business peacefully until the situation clarifies. You English have bought us time.'

'Why are you so opposed to Barr? Is not your cousin the Earl's son-in-law, and heir to the Earl's immense fortune and lands, some of the benefits of which in the Scottish tradition would flow onto yourself?'

'Our cousin Alistair, or rather his father was a fool. The Earl outfoxed him by including so many codicils and exceptions in the marriage agreement that Alistair receives very little that he can use in his lifetime. He is virtually guardian for any child that he has, and we his relatives get nothing. The courts at the moment are examining the legalities of this situation. To be realistic as long as Barr survives in any position of power we are not likely to receive a favourable decision. What makes it worse Alistair is not supporting us. He seems completely dominated by his wife. Now gentlemen why are you here?'

'To discuss Alistair. Some weeks ago Alistair finished up a prisoner or guest in Castle Clarke on the west coast. A joint Scots–English mission was sent to free the prisoners. We were successful, but now we cannot separate villain from hero. We are loath to release anybody until we understand why and by whom they were kidnapped. Would Alistair carry out orders from the Earl of Barr, or would he be out to undermine any such instructions?'

'All of us Stewarts fought for the old King and rallied to his son. To our surprise the Earl offered his daughter in marriage to the family. Our uncle, Alistair's father was flattered and readily accepted the offer believing it might bring the Earl of Barr and the King's supporters together. Until the

Earl's fortunes began to falter Alistair believed he would get more from the Earl by being his dutiful son- in-law. In the early days he would never bite the hand that fed him. It grieves me to say that Alistair has been a trimmer in recent years, although of late I hear he has dedicated himself to serving the King, with whom he appears to have developed a close relationship. The last I heard of him he was a courtier at Stirling Castle. Perhaps he is coming to his senses at last.'

'His problems are about to escalate. He has sired a soon- to-be- born heir for the Earl,' revealed Andrew.

'Poor Alistair. If that child is a male he will be taken away from Alistair and brought up as member of the Earl's family. His father's role and family will be relegated to that of the occasional visitor, if that. Whoever kidnapped Lady Elspeth did so either to prevent such an eventuality or, if Barr ordered it himself, to effect such a plan.

If Barr is behind this kidnapping then Alistair's life is certainly in danger,'

'Well he has managed to avoid the Earl's men so far and is under our protection,' said Luke.

'Don't fool yourself Colonel. If Alistair is in your castle the Earl of Barr's men are also there–awaiting their chance to kill him.'

10

Andrew interrupted, 'That's a worry. The Scottish troops assisting us are from the Earl of Barr's own regiment.'

'The Earl is no fool. He works at different levels. His troops will keep everything under control, but they would not break the law. This is a moralistic regime. No, Barr will have another agent, a secret agent, ready to kill Alistair—an agent he can disown or fatally discard.'

Luke, as he was about to leave, asked whether there were members of the Barr family living locally who were critical of the Earl. Alistair's cousin answered, 'The Master of Ochilmeath, the Earl's estranged younger brother. Follow the path along the river for two hours. The man is rarely sober, but if you are an enemy of his brother he will be your friend. Your Scottish companion will be even more unwelcome there.'

It was dark as two English soldiers and a civilian approached the decaying walls of Ochilmeath castle. Their previous host had reluctantly provided some clothes for an irate Dugall. Opposite the castle was Struthers, an inn, which unlike its English counterparts was for travellers, not imbibers. This was deep inside Kirk dominated territory and all landlords were models of moral rectitude. The local example, a corpulent middle-aged man with a large belly, asked gleefully, 'Are you Englishmen lost? Usually you return to the large towns before dark.'

Luke replied tersely, 'We are not lost. We have business with the Master of Ochilmeath in the morning.'

'That old reprobate! Take him away with you for his own good.'

Luke observed to the landlord how well the English were accepted locally. 'Don't be fooled Colonel. Scotland is on the brink. It has no effective government. It is divided in two, as is the Kirk. For the moment English occupation spares us the obligation to take sides. The English maintain law and order which supports trade. Once we have a united nation the focus will be on your eviction. The bottom line is clear. A republic is anathema to God.'

Next morning as they entered the castle's grounds Dugall and Luke argued. Luke wanted Dugall to wait outside and look after the horses while he and Andrew interviewed the nobleman. Dugall refused. As he represented the Earl of Barr he insisted that he should be present to counter any lies the drunken lord might advance. Andrew suggested a compromise. Dugall should be with them, but should not ask any questions. He could hear what Ochilmeath said, and give his responses after they left the castle.

Despite the general air of decay the part of the castle which was inhabited was weather tight and cosy. And the Master still had enough servants to minister to his needs. The servant who opened the door was amazed to find two English soldiers and a civilian, and was not quite sure how to react.

'We come in peace. We wish to speak to the Master of Ochilmeath about his niece Lady Elspeth Stewart.' Luke unknowingly had hit a soft spot. Ochilmeath had been very close to Elspeth. Luke explained that she had been kidnapped and that his unit had rescued her but was at a loss as what to do next. Some said she was safest under English protection. Some did not trust her with her husband, others that she must be protected from her father. It was not surprising that Ochilmeath took up the last viewpoint.

Luke conveniently did not mention that the co-protectors were the regiment of the Master's hated brother.

'My niece must be kept away from her father. Whoever kidnapped her deserves a reward. The Earl of Barr is a cold heartless man who treats his daughter as he treated his wife. When his wife delivered a child, Elspeth, and died in the act, my brother was overcome for months. No, not with grief over the loss of his wife, who had once been betrothed to me, but by her failure to deliver him a son. From the moment of her birth Elspeth had one role in her father's eyes. She had to be married off as soon as possible and produce a male heir. If he had no son, he would at least have a grandson.'

'That situation is nigh approaching. Lady Elspeth is well advanced. The Earl will soon have his grandchild,' said Andrew.

'Doesn't all this ignore the role of Elspeth's husband who would inherit the Barr lands on the Earl's death?' Luke questioned.

'It's a complex legal matter which could be solved very easily if Elspeth delivers a son, and Alistair dies before the Earl. The grandson would take it all. I am thirsty. Join me in a drink.'

Beakers of a spirit were produced and the four men indulged for some time. Luke recognised a familiar taste. 'My lord how is it that a Scottish nobleman has Irish whiskey?'

'During the Civil War this castle was garrisoned by Irish Royalists. The officers brought wagonloads of whiskey with them. When the armies of my brother relieved the castle in the name of the Kirk I helped myself to the whiskey, and hid it from the troops of the Scottish Parliament. It has kept me happy ever since.'

It kept them all happy for the rest of the day. As they prepared to leave next morning the Master of Ochilmeath made a final contribution. 'My brother, that righteous god-fearing leader of the realm, has a man who does his dirty work. He was sent here to kill me, but my retainers thwarted him.'

'What does he look like?'

'No one knows. He is a master of disguise. He dresses like a peasant, a great noblemen or a cleric. He is sometimes clean shaven, sometimes with a beard, sometimes his hair is long, sometimes cut shorter than your own.'

'Is it one man or several?'

'My brother would not place his own safety at risk by making more than one man privy to his secrets. Such a man may already be at your Castle Clarke, sent there to kill one of his enemies. Sir Alistair would be an obvious target.'

Only Aiden Mackelvie's lands remained within easy reach of the travelling inquisitors. They turned north into the foothills, and then through a series of glens and mountain passes until they reached the laird's house. Mackelvie's steward explained that his master had disappeared several weeks ago, and nobody had heard anything since. Luke indicated that they had news of the laird's fate and wished to speak to his wife. In a few minutes a young vivacious woman arrived. 'Gentlemen, I am Violet Mackelvie. You have news of Aiden?'

Violet Mackelvie was tiny. She was well under five foot tall but to Luke's experienced eye perfectly proportioned. She had light brown hair and a peaches and cream complexion. Luke was reminded of his sister's childhood doll. He replied, 'Yes, we belong to a joint English-Scots mission which has located and rescued Mr Mackelvie from his kidnappers. He is being held at Castle Clarke not very far north of here,' answered Luke.

'I have never been to Castle Clarke, but I know Lady Fenella very well. We spent a lot of time together in earlier times. Why would they kidnap Aiden?'

'We are not sure who actually kidnapped Mr Mackelvie, or if he was kidnapped at all. We hoped you might be able to help us,' Luke confessed.

'Aiden is not popular with the inhabitants of the glens to the north. Many lost their land to imported workers and farmers This was once Cameron territory, but there has been a progressive infiltration of Campbells. This conflict between them flares up frequently in mutual raids on each other's property but I can't see how kidnapping Aiden would achieve anything. In the end people would be wary of provoking the Marquis of Argyle by attacking his major local representative.'

'Did anything unusual happen in the weeks leading up to his disappearance?'

'Aiden was distressed over the conflict between the Marquis and the Marquis's eldest son Lord Lorne. The son wanted the Campbells in this area, including Aiden, to come out in support of the King, and demand that he be made the real head of state and commander of the army. The Marquis was opposed to any such declaration, and was working with his friend the Earl of Barr to reach a compromise–giving the King some power, but keeping real control in their own hands.'

'Do you have links with the Earl of Barr?' asked Andrew.

'Yes, the Earl is my patron, and it was he who arranged for me to marry Aiden.'

'Did any one visit Mr Mackelvie after which he appeared strange or worried?'

'He was accosted on the road by a gang of brigands who did not harm him personally nor take any of his property, but he had, according to one of the servants, a shouting match with the gang's leader.'

Luke asked to see the servant who had been present at his master's confrontation with the brigands. The servant dismissed the event, 'It was not a major disruption. The master had some twenty horsemen with him and the brigands had four.'

'So who initiated the meeting?'

'The brigands rode towards us, and shouted from some distance that they had a message for Mr Mackelvie. The master accompanied by me left our main group and rode to the brigands. Their leader took the master's reins and the two of them moved out of my hearing.'

'And how did your master react during the conversation?'

'He was anxious, and the brigand became increasingly angry. Both raised their voices.'

'And what did you hear?'

'The brigand told my master to keep his promise, and act on it. The master replied that times had changed and he would not. He then signalled for me to join him, and we rode back to our group.'

Luke dismissed the servant and asked Violet, 'Was there anything about the day he disappeared that was out of the ordinary?'

'Yes, he left behind his favourite white stallion. Aiden never travelled anywhere without that trusty steed. His affection for that beast had become a bit of a joke amongst the servants,' Violet replied with a tinge of sarcasm. She indicated that she would answer no more questions, and invited them to eat with her. Over a substantial meal, and much to drink everybody relaxed.

Violet asked, 'Do not treat me as a jealous woman, but have you noticed any special relationship between my husband and Lady Fenella?'

'Lady Fenella has a special relationship with many of the males, but I have never seen her with Mr Mackelvie,' replied Luke. 'Why do you ask? Does your husband have a history of chasing other women?'

'Oh no, since we have been married Aiden has been a devoted and loving husband. But he had a history with Fenella. For years Aiden was betrothed to Fenella, but his family became tired of waiting and the Earl of Barr intervened and within weeks I was married to Aiden instead of Fenella. A short time later Fenella was married to a much older laird from further up the coast.'

'Did Aiden see much of Fenella during their betrothal, or was it an arranged engagement in which neither party met the other?'

'They saw each other often, and were close friends. They wanted to marry.'

'So what was the delay?'

'Aiden was related to the Marquis of Argyle and his family did not think Fenella was a suitable match.'

'So Aiden's family broke up a love match, and married Aiden to yourself as part of family politics.'

'That sounds horrible Colonel, but it is the truth. But after our marriage we fell in love with each other.'

'Mistress, think back to the day Aiden was confronted by the brigands. What was his mood when he returned home? Did he say anything that might throw light on that meeting?'

'He was very quiet, but at one point he threw his goblet across the room and cursed.'

'What did he say? What did he curse?'

'That was the strange part. He cursed a thistle.'

11

'**W**as he more specific?' asked an animated Luke.

'He cursed black thistles.'

The soldiers were dumbfounded. Violet noticed the reaction.

'Important?' she asked.

'We suspect your husband was abducted by a group called the Black Thistle.'

Violet did not react, and deliberately changed the topic, 'When will Aiden be coming home?'

'Until we find out more about the situation it is not be safe to release him,' Luke replied.

'He will be safer with me than locked up with Fenella,' said Violet, with a sudden flash of fire.

'Maybe, tell us more about Lady Fenella. How is she dangerous to your husband? ' asked Andrew, sensing that her rising antagonism towards the mistress of Clarke Castle might reveal useful information.

Violet remained silent for some time but finally commented,

'I know Fenella very, very well. For two years Fenella and I lived together as gentlewomen companions to the younger daughters of Lord Montstone. My father believed that I would enhance my marriage prospects by being in Edinburgh close to important people. Fenella was in the same position. Her father, a diplomat, was always away on state business. Her mother died when she was very young.'

'Your father was obviously right. You have done well but at Fenella's cost.' Luke blandly observed.

Andrew however cut to the quick, 'The Earl of Barr negotiated Fenella's initial betrothal and her replacement by you. Why did the Earl of Barr decide Fenella was not a suitable match for Aiden Mackelvie?'

'How does this help you solve Aiden's kidnapping?' asked Violet, reacting to the increasing personal nature of the questioning.

'Maybe Fenella was behind it. Why did the Earl of Barr change his mind?' repeated Luke.

'If you have spent any time with Fenella you would know. She took several lovers while at the Montstones.'

'Why did Lord Montstone allow this to happen?'

'Montstone was himself involved. Lady Montstone discovered what was happening and informed her brother, the Earl of Barr who rejected Fenella as a suitable bride for a Campbell, and ejected her from Montstone's residence. My betrothal to Aiden served two purposes. It gave him a suitable wife, and it removed me from the immoral influences that everybody now assumed dominated Montstone's household. Fenella, a major embarrassment to the Kirk- loving elite, was found a husband on the edge of civilised society, the nefarious Sir Derek.'

'Do you know Sir Derek?'

'No, but I hear he will do anything for money. He is a very wealthy man, but spurned by decent society.'

'Why spurned by society?'

'There are rumours that his retainers are a violent lot of brigands who terrorise the smaller farmers. The Marquis of Argyle asked Aiden to take some action to protect Campbell settlers in the area from attacks allegedly mounted by Sir Derek.'

'So Derek had a reason to kidnap Aiden?'

'Yes.'

'Let me return to Fenella at Montstone's. Did you recognise any of her visitors?'

'Not by name. There were several men who wore a uniform similar to your companion.'

Dugall who had discarded his civilian disguise on leaving Ochilmeath blushed, 'The Earl of Barr's regiment has several hundred men.'

'It was a senior officer, Captain,' Violet responded carefully scrutinising the embarrassed Scot.

'Anybody else?' asked Andrew.

'I didn't recognize anybody, but Fenella told me just after the storm broke that her most constant companion was innocent. He had been a cover for someone else whose identity would rock the government, and certainly embarrass the Earl of Barr.'

'Was it the Earl himself?' asked Luke

'No. She specifically denied that. And I believe her. The Earl of Barr is a practicing puritan, and no hypocrite. And he is ruthless. If Fenella had been his lover she would not be allowed to roam freely in the Western Highlands with such a secret. For a King to have a stream of lovers is acceptable, for Barr even a whiff of such gossip would be politically fatal.'

'But if there was any truth in such a rumour could the kidnapping of all those people be a cover to remove Lady Fenella?' Luke probed.

Andrew whistled through his teeth, 'Cuds-me! Fenella the victim not the guard!'

'Fenella a victim? That would be a first! Gentlemen you may stay the night on the rugs in front of the fire. In the morning I will accompany you to Castle Clarke and reclaim my husband,' announced Violet.

'No Mistress, given your past with Lady Fenella it will complicate our enquiries. We will sleep in your barn, and leave before dawn for Castle Clarke.' Dugall excused himself and went to check on their horses and supplies.

As if freed by Dugall's departure Violet came close to Luke and took his hand and as a tear ran down her cheek she whispered,

'Please Colonel, I must come with you to tell Aiden personally that his brother, Angus, has been murdered. We must arrange for the funeral.'

'Murdered! What happened?' asked Luke not releasing Violet's hand.

'His decapitated body was found in a churchyard opposite a notorious drinking place in Oban. His torso was placed between two tombstones, and his head stuck on one of those spikes that surround the graves of the wealthy.'

Luke took in a deep breath, 'That is a strange place for a gentleman to be found.'

'Even stranger was that he had discarded his outer garments and was dressed like a vagabond.'

'What relationship did Aiden have with Angus?'

'They were political allies His younger brother was an agent of the Earl of Barr and arrived here a week or so ago on the Earl's business. He was heading for Oban to deliver the Earl's instructions to somebody.' Luke gave Violet a big hug and with Andrew retreated to the barn. Luke nevertheless refused Violet's request to accompany them the next morning when the soldiers left for Castle Clarke. With Dugall out of hearing, Luke and Andrew discussed the implications of her evidence. Was David involved in the murder of the Earl of Barr's agent? Was Mackelvie's brother the designated contact whose position David had usurped? Were they dupes of a complicated Scottish conspiracy?

On return to the Castle Luke briefed David, except for Mistress Mackelvie's evidence, and indicated that he would leave in the morning for the far north to investigate Janet Hudson and Mungo Macdonald. David was not enthusiastic. 'No Luke, the little information you will gather is not worth the time, the expense or the risk that such an enterprise will take. Call it off. If you go overland up the Great Glen towards Inverness you face fanatical royalist clansmen who hate us as much as they hate you.'

'Is there a safer way?'

'By sea around the north of Scotland, and then down the east coast,' David replied.

'Cromwell would not divert a ship for such a purpose,' answered Luke.

'You have authorization from the General to act in any way necessary to complete the mission,' countered David.

'Yes, but the General would expect it to be used only in critical circumstances. Investigating the family history of two prisoners hardly meets the unwritten criteria.'

Next day an armed English frigate arrived at Castle Clarke to deliver ammunition and additional cannon. Luke spoke to its captain who absolutely rejected Luke's request to take his party through the treacherous seas to the north. He claimed his orders from Cromwell were more recent than Luke's and therefore took priority. Just as Luke was about to leave, half angry and half frustrated, the captain offered him a lifeline. 'We passed a large, slow merchantman that is delivering supplies up the western coast. It should be here some time tomorrow.'

Two days later the merchantman sailed up the loch. Its captain had goods to deliver to the outer isles before he turned east, but for a price was willing to take Luke and his men to Inverness. The following morning Luke, Andrew, Dugall and five English troopers boarded the ship. The swell was considerable and the ship rolled from side to side in the most alarming fashion. Luke was seasick.

Later that day the boredom was relieved when a crewman shouted,

'Sail to the starboard.' Luke, although still nauseous, struggled onto the deck and recognised that the large frigate flying the standard of St Patrick was indeed a heavily gunned Royalist warship. The King had used his depleted fleet to stop English supplies reaching Ireland. Now he was extending this policy to hamper English activities in Scotland. The frigate was soon alongside, and a cannonade began. The merchantman was quickly de-masted, and damage to its rudder left it floating at the whim of the elements.

The frigate's speed took it past the wallowing merchantman. Its captain's intentions were clear. He would come around and finish off the disabled merchantman. Luke and his men gathered on the deck having rescued their horses from the fast filling hold. The frigate was beginning to turn when it disappeared. A thick fog had descended. Darkness soon followed. The master was blunt. With luck they would drift onto a sandy beach, without it the ship would be smashed to pieces on the jagged rocks that infested the offshore waters.

No one slept. The swell created a cacophony of sounds amplified as more of the damaged mast and its rigging fell to the deck. The master ascertained that the ship was floating backwards and placed a sailor in the stern to monitor the depth. Very soon he was calling depths of ten feet, then nine. The ship was soon aground. In the darkness no one could make out the landscape. Then it happened. There was a cracking sound and then an immense shudder as the ship began to spin ever so slowly in a ninety-degree angle. Nobody could swim. The ship began to sink as water poured in where it had been ripped apart by the rocks. The deck sank just below the water but no further. It was stuck in the sand. They would not drown, but within hours should the weather deteriorate the ship would disintegrate. The master advised that everybody stay where they were until daylight, when they would be able to assess their exact situation.

Dawn revealed that the ship had hit rocks a few yards off shore where a long sandy beach stretched out as far as the eye could see. Luke jumped overboard, and was delighted that the water was only a few feet deep. Luke ordered his men to lead their horses ashore. The sailors quickly followed but the master and his boatswain stayed on board gathering together as much of the cargo that they could save. The soldiers made it to the shore and temporarily freed their horses to feed on the lush grasses that grew up to the edge of the sand.

The Royalist frigate reappeared and slowly moved towards the beached merchantman. This time it reduced sail, dropped anchor, and blasted into oblivion the battered ship and its remnant officer crew. Luke's men, and the sailors moved inland out of range of the ship's cannons. Once the frigate sailed away they returned to the beach, and waded into the water looking for food and valuable debris. Barrels of liquor floated in on the tide. The soldiers waded out to what was left of the ship where they found large pieces of salted meat caught amongst other flotsam. Luke was not sure however whether he was treading on salted meat, or dismembered crew.

Luke saw that some barrels floated lower in the water than others. The markings were familiar. These were barrels of gunpowder produced for the English government. They should have been unloaded at Castle Clarke. The master was obviously retaining military supplies and selling them off to the highest bidder, probably the Royalist army of Highlanders gathering near Inverness. A fire was lit and the men dried their clothes, and warmed themselves for several hours. The rescued meat was cooked and eaten. Luke asked the sailors where they were. Their spokesman suggested, in a broad Northern English accent, that they had made little progress. Once disabled they had drifted back many leagues towards the point of their departure. The mountains that rose behind them were not far north of Castle Clarke.

The troubled sailors gathered around Luke. A passive request soon escalated into a belligerent chant, 'Take us back to Castle Clarke!'

12

Luke would not be dictated to by an unruly mob of sailors. Andrew saw his predicament and with the other troopers rode into the gathering, firing their carbines into the air. The sailors dispersed and after some time their spokesman reapproached Luke. 'Colonel, my comrades are terrified. They fear they will be left at the mercy of the barbarian highlanders. They are southern English, afraid of their own shadows. They may attempt to disarm you, and steal your horses.'

Luke had not expected this mutinous reaction from the sailors. He could escort them back through Cameron territory to Castle Clarke, augment his group and then reach Inverness overland along the Great Glen. Andrew opposed any suggestion of delay. 'If we waste any time Scottish politics will have moved on and this whole mission will be irrelevant. We cannot afford any delays. We must move directly to Inverness.'

Dugall agreed, 'No number of troops can guarantee our safe passage. Fifty troops will be no better than five. We will confront large numbers of clansmen moving towards Inverness to join a Royalist force that intends to place Charles at the head of a united Scottish army.'

Luke accepted Andrew's point, and Dugall's further suggestion that the sailors be taken up the Great Glen to Inverness where they could find new employment. The soldiers realised that the sailors would not willingly agree. Luke lied. He told the sailors that they were being escorted back to Castle Clarke and provided several barrels of strong liquor for the sailors to celebrate this popular decision. Not a single soldier drank. Well into the night, with the sailors either sound asleep, or in such a drunken stupor that they could not retaliate, the soldiers removed every weapon they could find. Next morning after an hour into their journey Luke informed the sailors that they were walking to Inverness. One sailor protested and rushed at Dugall with a large piece of driftwood. Dugall shot him dead. There was no further resistance.

The party headed northeast along slender lochs, interconnected valleys and narrow mountain passes. This was a more settled area and every glen was well inhabited. They met no resistance. Instead they received traditional Highland hospitality. Dugall explained that these areas were well controlled by their clan leaders who were cautious about embarking on any premature attack on the English. Luke had made it clear at the first hamlet that he came in peace, and had a single aim to reach Inverness.

Inverness was almost an open city. Although the surrounding countryside was Royalist the city itself pledged loyalty to the existing Scottish government, and had a strong Kirk influence. The town was also not overtly opposed to the English as their warships appeared constantly off the coast and were resupplied by the town merchants.

On reaching the city Luke allowed the sailors to find their own way to the harbour, and eventually he found Hudson's house where a servant motioned for Dugall and Luke to follow him into a small antechamber. There a young, tall, sandy haired man introduced himself as John Hudson, Janet's brother, 'You have news of my sister?'

'Yes, we have rescued your sister, and have her kidnappers under guard,' Luke replied.

'When will she be home?'

'That is why we are here,' added Dugall. 'We are unable to separate the victims from the kidnappers, and we need to understand why the victims were taken. Why was your sister kidnapped?'

'Well it was not for her money. When father died the whole estate, landed and moveable came to me as the eldest male heir. Father left nothing to his daughter having prevailed on an old family friend to provide her with a dowry, and find her a husband.'

'Why would any old friend be so generous?'

'Thirty years ago two young Scottish cornets joined the French army to fight the Spaniards. In an engagement the friend was attacked by superior opposition and was badly injured. In fact he was left for dead. Just before the Spanish officer began his inspection of the wounded opponents to put them out of their misery father rode through the Spanish lines, scooped up his friend and escaped as a convenient fog descended on the battlefield. The friend took months to recover but never forgot father's actions.'

'Who is this friend?' Luke asked.

'He became the Earl of Barr.'

'That is certainly relevant,' exclaimed Dugall.

'Why?'

'Nearly all those at Castle Clark have some connection with the Earl,' Luke explained.

'I doubt that the Earl would be behind Janet's abduction. It is more likely to be an extreme supporter of the Kirk, one of the puritanical moralising lower aristocracy,' suggested John Hudson.

'What would suggest that?'

'My sister is very moralistic. She is an obsessive Kirk attending, Bible reading saint, the perfect woman for similarly minded aristocratic mothers looking for the perfect wife for their son and heir. The Earl of Barr's selection for a husband, pragmatic and political in his eyes, was personally somewhat wayward. This may have displeased many of the religious faction who hoped Barr would choose their offspring. He is giving Janet an immense dowry.'

'Did you deal directly with the Earl over these matters?'

'No, he sent another of father's old friends who is now one of the earl's secretaries to talk to me.'

'Did Janet meet him?'

'Yes.'

'Have you had any contact with the Earl since she was kidnapped?'

'Yes, but it was a bit unnerving and now I am not sure whether the man really was from the Earl. I sent a message to him that Janet had been abducted by a large group of horsemen last seen heading down the Great Glen. A week later a man arrived dressed like a clergyman, but carrying both a sword and dirk. He claimed he had been sent by the Earl to investigate the matter.'

'What was unnerving about him?'

'He accused me of being party to Janet's disappearance to prevent the Earl achieving his purpose for her. He pushed his face right into mine, and threatened that if I lied to him I would lie to nobody else.'

'Did he give you a name?'

'Yes, he said he was a Mungo Macdonald, but I knew he was lying.'

'Excellent!' muttered Luke.

'What have I said?'

'A Mungo Macdonald is also a prisoner at Castle Clarke,' Dugall replied.

'This man must have tracked Janet. That is what he said he would do. I have since heard that he beat up people who crossed him. I doubt if he is a cleric. And he is certainly not Mungo Macdonald.'

'Why not?'

'The parish church is next door. If you go into the churchyard the first tombstone, which you can't miss, is erected to the memory of our most recent minister, the Reverend Mungo Macdonald. The Earl's man could not have missed it. It provided him with a suitable alias.'

'Tell us about your late minister?'

John Hudson gave a short biography of his late parson. Luke groaned, 'Cuds me, the man we have at Castle Clarke is an impostor. The details you have given me are exactly those which the man at Castle Clarke gave to one of my men. We came north to find out more about the Reverend Macdonald. That project is at an end. Why would the Earl send such an unsavoury man to investigate Janet's disappearance?'

'The Earl has always met fire with fire. If brigands took Janet he would send similar types to deal with them.'

Dugall asked, 'Where do you and Janet stand in the current political turmoil?'

'Like my father before me I believe that Kings are appointed by God and therefore the King must respect the Kirk, and the Kirk must respect the King. When the King does not respect the church enough, as is the current situation we must maintain the reign of the King, but limit his real power until he accepts without question the teachings of the Scottish church.'

'You have said that Janet would make the perfect wife for one of the Earl's more conservative political and religious allies—a perfect God-fearing woman, and politically correct. In that case who would benefit most by Janet's abduction?'

'No one would gain. There can only be losers. It must be embarrassing to the Earl at a time when he is politically vulnerable,' suggested John.

The ringing of bells interrupted the conversation. 'Don't be alarmed. Either an army of Highlanders have come down from the hills, or a squadron of English warships are in the firth.'

'So which is it?' asked Luke.

'It doesn't matter. It's a call for the town's garrison to report to barracks.'

A servant informed his master that there were English ships approaching the harbour. Luke asked, 'Do the English put troops ashore?'

'Yes, while fresh supplies are loaded, they often reconnoitre the hinterland for signs of Highlander activity. The local merchants who provision the English ships also supply your navy with valuable information.' Luke thanked John Hudson and sought out the commanding officer of the English landing party with whom he negotiated passage for his men and horses, although they were dispersed amongst several of the English ships. Luke and Andrew were surprisingly allotted to the same ship, and were soon taking a drink with its naval captain. The squadron's orders were to sail north around the extremities of Scotland and down the west coast between the mainland and the isles to seek out Royalists ships that had been attacking merchantmen.

The squadron headed down Moray Firth into a bleak North Sea. As visibility lessened the captain reduced sail and Luke felt that their ship was falling behind the rest of the squadron. Next morning the ship battled alone westward around the north of Scotland. The captain did not seem to be worried that he had lost the rest of the fleet. As the day progressed he continued further west than Luke thought appropriate. The captain was sailing to the outer isles. When Luke complained that the squadron's orders were to sail down the west coast of the Scottish mainland direct for Oban and Castle Clarke the captain claimed he had independent orders to search the outer islands for signs of Highland troops. Then a sailor who Luke recognized as one from the shipwreck jeered at Luke and ran his hand across his throat. Something was amiss.

On reaching one of the outer islands the ship entered a small but deep bay, where riding at anchor was the Royalist man of war that had attacked their merchantman earlier that month. Luke turned towards the captain expecting orders would have been given to man the guns, and bring the ship into a position to attack the stationary enemy. Instead two officers with pistols drawn confronted him, the cross of St George was lowered, and the royal standard raised in its place. Another of the republic's ships was deserting to the Royalists.

'What happens to us?' asked Luke.

'You will be left to rot on this god-forsaken island. It is almost deserted. Our sister ship has taken on board most of the local clansmen to augment the Royalist army in the north. Within the week they will unite with the Scottish army at Stirling and Charles II will drive Cromwell from the land,' announced the naval captain.

'Sir, you ignore simple facts. No matter how large an army can be put against Cromwell, it will fail. Amateur barbarians against the discipline of the New Model!' boasted Andrew. He was clubbed to the ground with the butt of a musket.

The hovels on the island were not fit for animals. The women and children had been moved to a larger island and all the able bodied men conscripted into the Royal service. Only a few very old men remained. Luke and Andrew helped an octogenarian fisherman with his nets. Fortunately their exile was short. Within a few days the bay was filled with ships, the remainder of the English Republic's squadron that had left Inverness with Luke's men. A landing party included Dugall, who was delighted to see Andrew and Luke. Dugall explained that the commander of the squadron had intelligence that the captain of Luke's ship would defect. When the ship did not attempt to catch up to the squadron the commander retraced his course and unexpectedly came across the missing ship accompanied by a Royalist frigate midway to the mainland. Both were overloaded with armed highlanders and were easy prey for the English warships. Dugall completed his report. 'One of our naval officers recognised that the clansmen came from this island. We escorted the enemy ships to the other side of the island, unloaded the local highlanders, formally seized the two ships, conscripted the Royalist seamen, and executed their officers.'

13

The day after Luke, Andrew and Dugall returned to Castle Clarke they reviewed the situation with David and Harry. In their absence David had hardened his position. He would not confide in his English partners. He would tell Luke nothing of his meeting with the King and his mission to the woodcutter's hut. He would continue to conceal his recognition of the voices of Sir Derek Clarke, Aiden Mackelvie and Duff Mackail. He did elaborate in more detail that the knowledge of the plot to kill Scottish nobles and English generals stemmed initially from intelligence reported by Barr's agents. He explained that the Earl of Barr had sent him personally to meet General Cromwell, and negotiate their joint mission.

This limited confession did not fool Luke. It only confirmed his growing mistrust. How could anyone be sure that the latest revelations were in fact the truth or the whole story? But for the moment Luke would reciprocate David's outward cordiality. In reality mutual distrust dominated. Luke too could keep secrets. He would not let David know that he knew that the decapitated victim was no vagabond, but Barr's original contact.

The officers did pool the information they had on the inhabitants of Castle Clarke. Sir Derek was a man who acted through monetary self-interest—a weakness that they may be able to turn to their advantage. Luke argued that the critical person in their investigation was Lady Fenella. 'Neither Mackelvie nor Fenella have been truthful in what they had told us. Fenella probably holds a grudge against Mackelvie for deserting her, and hates the Earl of Barr for organising her rapid change of fortune. And who were her mysterious lovers at Montstones? Do they have any part to play in the current run of events?'

David surprisingly agreed, 'Fenella's generosity towards men creates all sorts of problems and possibilities. Male reaction to her behaviour could be more central to understanding what is happening here than the high politics of Scotland.' Luke did not reply but wondered how far David's past relationship, if any, with Fenella was a critical factor in the situation.

Harry was not as diplomatic. 'Were you, David, the senior officer from the Earl of Barr's regiment who was Fenella's lover?'

'A contemptuous question!' responded Dugall defending the honour of the regiment. 'Lady Fenella's lovers did not include senior officers of the regiment. I know all of them. They are honourable men. Her lover donned such an uniform to move freely in and out of Lord Montstone's house.'

Harry excitedly continued, 'So what are you suggesting Dugall? That Fenella's lover could be anybody except a member of your regiment. That's hardly logical.'

Luke intervened, 'Dugall has a point. If you were secretly having an affair with a notorious woman in a household connected to the Earl of Barr it would be politic to disguise yourself as a member of the Earl's regiment. Your presence in the household would be less conspicuous.'

Dugall continued, 'It would not surprise me if the seducer was some Royalist blade, maybe the King himself.'

Luke smiled, while David glared at Dugall.

'Rubbish!' said Andrew.' Young Charles is a lusty lad, but he is no fool. Even the slightest whiff of such a scandal would have been used by Barr's faction to denigrate the boy, and destroy his growing support amongst the conservative Scots.'

'Cuds me Andrew, you are a prude. When has a lively sexual appetite and a string of lovers ever negatively affected the reputation of a King?' replied David.

Harry summed up, 'So Lady Fenella is not an innocent bystander in whatever is going on, but possibly a major player who needs to be watched carefully. It would be unwise for any of us to get too deeply involved with the lady.' He looked directly at Luke, who averted his eyes. Harry, from family of Puritans, was often a moralistic prig where his colonel's lusting after women was concerned.

Dugall disagreed with the emphasis on Fenella. In his eyes the critical person in the investigation was Duff Mackail. All admitted that they knew hardly anything about him, and that it had been a mistake to allow him and his men to leave the Castle. The officers continued their review. Andrew suggested that the marriage of Sir Alistair and Lady Elspeth and the imminent birth of an heir should be central to their investigation. Was Elspeth politically with her father or her husband? If forced to chose whether her husband and his family, or her father should control the upbringing of the child she carries, which way would she go?

David agreed that the political implications of this coming birth should not be underestimated. 'Elspeth's abduction has much to do with controlling the future of her child. Alistair's late arrival on the scene could either be as an agent of his father-in-law—most unlikely—or on behalf of the Stewarts. Either family could be behind her abduction.'

'That suggests that both parties have an agent here to protect their interests. If Alistair is here to protect the Stewarts, who acts for the Earl of Barr?' asked Andrew.

'Is that your role Dugall?' Harry shouted across the table.

The Scots captain bristled at Harry's innuendo, but was saved by David's intervention. 'We are here to protect the interests of the Scottish government which at the moment is administered by the Earl of Barr. You Lieutenant obey General Cromwell, who is simply the commander of English forces in Scotland. You are not responsible as far as I can see to the English government in London.

You are on shakier legal grounds than we are. We act for the Scottish government, you for a faction of the English military.'

Luke returned the discussion to the review of inhabitants.

'Caddell is easier to understand, and his enemies are numerous. Anybody here could be related to a victim of the witch-hunts. For example are any of the inhabitants of this castle related to Lady Dalmabass'?

David added, 'It does not have to be Lady Dalmabass. Caddell as a witch hunter made enemies of hundreds of people.'

'Lady Dalmabass was rescued by her brother. Given what we know about relative ages Sir Alistair, Duff Mackail, Mr Mackelvie, Major Burns or Captain Sinclair could be that brother,' added Harry deliberately and provocatively.

Luke had had enough, 'Lieutenant stop assailing our allies! One more attack on our comrades and you will return to Ireland.'

The niggling Harry was unstoppable, 'Colonel I mean no insult to Major Burns or Captain Sinclair personally, but I point out that we are at war with Scotland. They have insulted the English Republic by proclaiming the son of the late tyrant as their King. Within a few months we will be at each other's throats again. I caution against a friendly relationship. Neither side should reveal to the other any matters that might affect the security of their state, or of the imminent campaign.'

'You forget Harry that we are under direct orders from the Lord General to undertake this mission with the Scots. We have a common enterprise, to uncover The Black Thistle, and to save the lives of our political leaders. What is happening here is a vital part of solving those problems. David and I are not naïve idealists. We both know when to end the alliance,' proclaimed Luke with an air of authority.

'What about Janet Hudson?' asked David turning the discussion onto neutral ground.

'That can only be an anti-Barr abduction. She has no assets of her own but those that Barr will bestow on her future husband. Her moral fortitude makes her a most acceptable wife for the God fearing, Kirk supporting gentry and aristocracy. She would not appeal to any but a narrow set of the Scottish landed gentry. An alliance with Barr would have been a major asset for members of this group, but as he is now on the way out Hudson's abduction loses any point. She is one person I feel we could send home to her brother. He will probably have to find her future husband much lower down the social scale,' concluded Luke.

Harry continued his negative approach. 'Perhaps Barr is behind the kidnapping so that he can get out of his promise to the girl's father?'

David spoke quietly, 'If any of the guests is in immediate danger it is Caddell as a hated witch hunter, or young Morag Ritchie. Ritchie's inheritance makes her a long-term asset for any of the great nobles to control. Getting her out of the Barr orbit may be behind her abduction, but there could be a more dastardly reason. If the murderer of her parents learnt that the girl's memory was returning, he would wish to kill her before she exposed him.'

'That would explain the attempt to kill her,' Andrew commented.

'The murderer would have to be very close to the girl to know this.' said Luke.

'Precisely my point. Who amongst our guests knows Morag or her family that intimately?' asked David.

'Place her under special guard!' advocated Dugall.

Harry, as if he had to speak whenever Dugall opened his mouth, commented, 'Is it not a worry that Lady Fenella and Lady Elspeth, have taken a special interest in the young girl? She could have confided in either one or both and their ladyships have passed on the information, maybe unsuspectingly, to the potential murderer.'

'Simply maternal instinct coming to the fore–at least as far as Elspeth is concerned.' stated Luke. 'How do we view Petrie?'

'Sir Malcolm Petrie has a distinguished career as a diplomat and lawyer. Most recently he has travelled the continent where he negotiated with all of the factions involved in Scotland. On behalf of the current government he persuaded the King to return to Scotland. Since the King has arrived has Petrie developed a closer relationship? Is he still loyal to the Kirk dominated government?' reflected Dugall.

'Petrie is a man of the world who would know exactly when to change sides. He is a typical political trimmer. I do not trust him,' Luke commented with some bitterness. 'Mackelvie is of the same ilk.'

'He and Lady Fenella have a past which might have annoyed not only her ladyship, but also Sir Derek or Duff Mackail. Aiden has clearly gone back on a commitment. His wife might suspect that his relationship with Fenella is still continuing. There are many disgruntled Camerons who resent this Campbell's expansionist tendencies,' said Andrew.

David concluded, 'That leaves Mungo Macdonald. We know that that is not his name. Is he an agent of the Earl of Barr sent to protect the missing women, or is he an agent of another whose agenda is anyone's guess?'

'He needs to be watched, but we will not give him any hint that we know he is an impostor. If he feels he has fooled us he might get very careless,' advised Dugall.

Luke summed up their review and announced, 'We have sufficient further evidence to re-examine everybody of note in the castle. I will begin in the morning without warning. I will start with Mackelvie and then Macdonald.'

14

Early the following morning a trooper woke Harry Lloyd, and told him somewhat incoherently that a household servant brought him a cup of mulled red wine the previous evening. The next thing he remembered was waking up, lying on the stone floor outside the chamber of Morag Ritchie where he was supposed to be on guard duty.

'What is worse sir,' continued the distraught soldier. 'Miss Morag is missing.'

'Good God, inform the Colonel immediately!'

'I can't. He is not in his room. I reported there first.'

'Jehu! What's happening?' exclaimed Harry.

'It's a disaster, Lieutenant. The sentries outside the chambers on my floor are missing or asleep.'

Harry and the trooper made their way to the lesser great hall–the barracks for the English troopers not on duty. Harry ordered a check on their charges and a search of the castle. John Halliwell immediately reported that Andrew Ford was also missing. His palliasse had been slept on, but his outer clothing was gone. Something must have awakened him during the night. The trooper who had raised the alarm tried to brighten the gloomy atmosphere, 'Perhaps the Colonel and Sergeant Ford are already on to the problem.'

'Trooper, sound your cornet! That should raise the dead.'

'Let's hope it doesn't have to,' muttered John only half jokingly.

Harry was surprised that he was not confronted by hordes of irate servants awakened earlier than their custom by the discordant blasts. He made his way to Sir Derek Clarke's room to alert him to the situation. The trooper assigned to the door was nowhere to be seen. Harry entered the room. Sir Derek was not there, and his bed had not been slept in. Harry made similar discoveries in Lady Fenella and Lady Elspeth's chambers. Half an hour later John reported, 'The castle is deserted except for two servants in the kitchen.'

'Did you question them about the disappearance?'

'They are lovers who hid themselves in a distant chamber to enjoy each other. They returned to their own quarters to find their companions had gone.'

'Our hosts and their guests are all missing?' asked an incredulous Harry of himself.

John reported optimistically, 'Yes, but the good news is that all our troopers are accounted for. They were drugged by the mulled wine. A few lay where they fell, but the rest were dragged into a small room off the great hall. The only English soldiers missing are Colonel Tremayne and Sergeant Ford.'

David was informed. He had more to add, 'The soldiers I placed at the entrance to the secret tunnel were missing when their relief arrived some fifteen minutes ago. It gets worse. Yesterday my men allowed a large number of carts, and dozens of horses to leave the castle. Sir Derek claimed they belonged to Captain Mackail. In fact they were your horses.'

Harry was furious. 'Damned Scots! Surely your men could tell the difference between a fine English cavalry beast and the piebald ponies that the locals ride? This has been a well-planned withdrawal. Whoever is behind this move, must have believed that our trip across Scotland had uncovered incriminating evidence. We must follow them immediately. As my men have no horses they can take over the defence of Clarke Castle. Major, you must lead the chase. If I can borrow one of your horses, I will come with you.'

'Certainly, but a small group of six will be sufficient, four of my men and ourselves,' replied David. 'Captain Sinclair, with Sergeant Halliwell as his deputy, will command the garrison in our absence.'

By mid morning the tracking party was ready. Two of David's troopers followed the secret tunnel and confirmed that numerous people had moved along it. A coif, handkerchief and bracelet had been dropped in the haste to escape. The tracks of many horses and a few vehicles that had been waiting at the end of the tunnel were still visible. On the first day the trackers followed Loch Linnhe to its head, then down its eastern shore and finally east along Loch Leven towards the high mountains. The following day they moved through these mountains reaching a high pass that was the only link between two glens. The way was deep in snow and they ploughed through intermittent snowstorms. As they moved down into the next glen David signalled for the troop to stop. Visible in the fresh snow were the hoof prints of numerous horses that had entered the main way from a different direction, but were now heading in the same direction as the emigrants from Castle Clarke. David feared that the fresh horse tracks would obliterate evidence of their quarry. Harry asked David where the Clarkes were heading. He shook his head. He confessed he was from the Borders and had little knowledge of the Highlands.

Towards the end of the second day they descended into a large glen. David confessed he had lost the trail. They would camp for the night and next morning begin their return to Castle Clarke. When the sun rose it lit up the face of a distant mountain. Harry exclaimed with delight, 'That must be them.' As the soldiers assessed the situation a group of locals were sighted further along the glen. David galloped up to them and engaged in a long discussion. He returned with a big smile on his face. 'They told me I was the second person to ask for directions into the mountains. Last evening a group of armed men whose leader the locals recognized as Captain Mackail had asked the same question. He had met with a large group consisting of many horses, people and vehicles that had had waited here for the captain to catch up with them. The locals were amazed that the large group left the traditional pathway to the next glen, and headed directly into the high mountains.'

'Let's get after them!' exclaimed an excited Harry. 'Did the locals mention a pair of strangers moving through their hamlet?'

'No, but if Luke and Andrew are following the Clarke's migration they would keep themselves hidden,' suggested David. 'Son, don't build up your hopes. Luke and Andrew are experienced soldiers, but they may be dead. The next possibility is that they are prisoners taken against their will. If they were free and following our quarry I am sure they would have left some clues for us to follow. There are none.'

After several minutes of silence David announced, 'At least we know where our quarry are heading. The locals revealed that the only edifice in that direction is a semi derelict but easily defensible tower house known as Greytower.'

Climbing the mountain was difficult, and the snowfall was little short of a blizzard, which obscured any distant view. By afternoon conditions cleared but shadows now hid the slopes on which they had earlier seen the group they were pursuing. The path narrowed and the troop was forced into single file. Rounding an acute corner David halted, and signalled his men to dismount. A small band of Highlanders was camped across the trail just ahead of them. They had a fire, and were preparing to eat from a large pot. The smell of a hot mutton stew drifted down the mountain path. David ordered his men back down the slope to a cave they had just passed. There they ate cold salted meat and hard biscuits.

During the night as the men snuggled together under their blankets to maximise heat, their horses suddenly erupted into human like screaming. Harry saw pairs of eyes darting between the distressed horses. A pack of wolves was attacking one of the horses. Harry fired his carbine at the leader of the pack. One of David's troopers did the same. The pack dispersed. Another of David's men who was familiar with the region shouted, 'Stop firing, you could have us killed!'

'How so?' asked Harry.

'Firing on these snow capped mountains can cause an avalanche and we could all be smothered by the moving ice and snow,' replied the anxious trooper.

David was cross, 'Of more immediate concern, is that the noise probably alerted Mackail's men further up the slope to our presence. For the moment calm your horses and maintain absolute silence.' The rest of the night passed peacefully, which worried David. As the sun rose David allowed his men to catch up on missing sleep. It was dangerous to proceed immediately. The enemy would have heard the musket shots, but it was too difficult for them to act during the night. They were waiting around the next bend. To mislead the Highlanders David sent two of his troopers back down the slope to wait in the large glen. A third trooper joined them on foot as the wolf attack had so badly mutilated his horse that it had to be killed.

It was early afternoon before David, Harry and the remaining trooper resumed their ascent of the mountain. Happily their opponents had tired of waiting for their prey, or may have even been confused by the troopers moving down the mountain. David and his party eventually saw, hard against a precipice on the next mountain, a single grey tower. David was decisive, 'There is no point continuing. We will return to Castle Clark and consider our position. If Luke and Andrew are in Greytower we will need a small army to rescue them. And if this weather continues a small army would not be able to reach them until the spring thaw.'

THREE DAYS EARLIER, CASTLE CLARKE

After meeting the returning Alistair, Luke informed David about the secret tunnel. He immediately posted two men to guard its external entrance. At dinner Luke was uneasy. The external signs appeared normal although Derek, Alistair and their respective wives were overly chatty, and unusually elated, while the rest of the dinner guests were sullen and hardly spoke. Perhaps Alistair's return had them worried. Luke wished he had chosen to eat with his men as had David and Harry.

The guests and Derek soon excused themselves, while Fenella guided Luke, Alistair and Elspeth to chairs in front of a blazing fire. The castle was cold, and Luke enjoyed the warmth of these giant fires, and the endless supply of whisky that appeared before him. Luke also enjoyed the company of Elspeth and Fenella. The former while attentive to her husband did not hesitate to flirt with Luke, and Fenella was exuding a sensuality that Luke found hard to ignore. After an hour Alistair and Elspeth begged their leave, leaving Luke and Fenella alone. She took Luke's hand and placed it provocatively on her breast. Within seconds Luke was gently running his hands over Fenella's body. Soon each was thrusting their tongues deep into the mouth of the other. Fenella whispered, 'Luke, follow me!' Luke followed her at a distance to the small staircase that lead to her room. One of Luke's troopers was already on guard duty. Fenella spoke to the trooper who momentarily moved down the corridor, and Luke took advantage of the trooper's distraction to slip into Fenella's chamber unnoticed.

The next hour was spent in an unbridled display of mutual passion that the moralistic Kirk would have condemned as satanic. Fenella led Luke into activities he did not know existed. After both were exhausted Fenella offered Luke her favourite drink. It was whisky that had been infused with the honey of the heather and various spices of which Luke recognized anise. As he drank he became unusually sleepy. The last thing he remembered was Fenella covering him with several layers of bedclothes.

While Luke was indulging himself his senior sergeant, Andrew, was completing his daily routine. He accounted for the troopers who were coming off duty as they retired for the night to the smaller of the great halls that had become their dormitory. He then checked throughout the castle that the troopers on guard were in position outside the doors of the senior household members and their guests. He then met his opposite number in the Scottish unit who was responsible for the external security of the castle at the main door of the castle, and mutually reported that all was well.

Well not quite. Due to the heavy snow the Scots sergeant had not been able to contact the soldiers sent to the entry of the secret tunnel. Andrew returned to his quarters. He passed a household servant dressed for outdoor activities, carrying a tray with steaming mugs upon it. Lady Fenella, given the coldness of the evening, had ordered that mulled wine be distributed to the troopers on duty. Andrew declined a mug, and once in his chamber built up the fire, removed his outer clothing, and pulled his blankets over his head. Sleep did not come. Something was nagging at the back of his mind. Why was the servant delivering the mulled wine dressed as if she was preparing to face a blizzard? And where were the other servants who after their superiors retired had a range of jobs to complete? The castle was too quiet.

15

ndrew tossed and turned. Eventually he rose, put on his cloak, and made his way to Luke's chamber. The trooper that should have been alert before the Colonel's door was lying on the floor snoring loudly. Andrew pushed open the door, calling on Luke to arouse himself. The fire in the room was out. Luke was not in his bed. It had not been slept in. Knowing Luke's proclivity for the sensual woman he ran to Lady Fenella's bedchamber. A familiar sight met his eyes, a trooper lying asleep outside the door, and the room itself empty. He turned to raise the alarm–but found his passage blocked by three tall hirsute men armed with thick curved swords. They bundled Andrew into Morag Ritchie's room, and through the hearth into a secret chamber, and beyond the second chimney, into the hidden tunnel.

The tunnel was alight with tapers, and quite noisy. Andrew could see many people ahead of him. Exiting the tunnel he was confronted by Sir Alistair. 'My apologies sergeant. You should have been fast asleep. Now that you are here, you can help.' Andrew looked around and could see a large number of people moving off into the darkness. Sir Alistair and half a dozen armed men were bringing up the rear. Then Andrew saw Luke. He was lying on a sled to which he was tied across his waist and by the feet. He was asleep, breathing heavily and snoring intermittently.

'Pull the sled carrying your Colonel! He will recover. He has a surfeit of whisky spiced with sleep inducing herbs.'

'Sir Alistair, what are you doing?' asked Andrew.

'I am not your enemy. I am moving the whole household to a more obscure and isolated place. The situation at Castle Clarke is heavily weighted towards the English, and the Black Thistle. You could reinforce it at any time with shiploads of soldiers, and your masters could have released the household despite Scottish objections.'

'Does David know what is happening?'

'No. He takes his orders from another group within our struggling government—a group who are a little too friendly with the English. I act for the King. In our new location no one can escape, and this includes the Black Thistle. No one will be able to send letters across the country, nor in any way effect their dastardly schemes. David's blockade of Castle Clarke was ineffective. As a sign that I wish to continue to co-operate with the English I brought Colonel Tremayne with me.'

'I don't think the Colonel would see being drugged and abducted as a manifestation of friendly cooperation.'

'The method may irritate him, but the result, exit from Castle Clarke, may serve him well. The atmosphere here is not conducive to his inquiry.'

'Who are these wild looking men that apprehended me?' asked Andrew.

'They are a small body of Duff Mackail's men who ensured that the inhabitants of Castle Clarke came with me. Mackail and the rest of his men will join us shortly.'

'How did you get the group to come with you?'

'I told them that they were about to be shipped out to England, and possibly on to Barbados. I also hinted that once out of English control they would all be freed.'

'How long can you sustain this move? Both the English and Scottish troops at Castle Clark will track you down.'

'Our new destination can be defended by a dozen men, and within a week or so would be impossible to reach. More frequent and heavier blizzards will block the mountain passes. We will have a perfect isolated environment in which to solve the mysteries we confront.' Alistair threw another blanket over Luke, and signalled Andrew to take up Luke's sled. He offered Andrew a thick cloak which the sergeant accepted with alacrity.

Next morning after they made their way along the snow free banks of an extensive loch Andrew could tell they were at last moving east—deeper into the mountains. Luke began to stir and eventually was on his feet with his restraints removed. Andrew filled him in with the details of what had happened. Alistair provided them with horses, which they instantly recognized as their own cavalry steeds.

Luke was elated. 'This could turn to our advantage. I was unhappy with the situation at Castle Clarke and particularly concerned about our ability to prevent information leaving the castle. Who knows? There may have been secret passages other than the one we uncovered. And above all we are free from David whose agenda increasingly is not ours.'

'So you don't want to escape? There are only a few Highland soldiers. We could easily outsmart them,' replied an arrogant Andrew, with all the prejudices of his race. His father had been a lowland Scot.

'No, Harry and David will eventually reach us. Anyhow Alistair could be a more trustworthy ally than David.'

'Isn't Alistair David's ally?'

'No, David claims to be working for the Earl of Barr, but in this time of flux all supporters of the Scottish government, especially the strong nationalists like David have to decide where Scotland should go in the future. Should they support the King, but declare an independent Scottish monarchy and cut all of his English ties; support the King in invading England and in the restoration of an English dominated British monarchy; or cooperate with our army in our liberation of the land? David is probably not sure where he stands; on the other hand Alistair has no doubts. He is totally committed to the King.'

'So what is his relationship with the Royalist Black Thistle?' asked Andrew.

'The Black Thistle appears to have developed an agenda of its own which Alistair and the King oppose. But how far any change of official policy relates to the murder of faction leaders and English generals we cannot be sure,' Luke surmised. All of a sudden there was a fracas further along the line of travellers and a person staggered away from group stumbling towards the edge of the loch. There were loud oaths in Gaelic, and a couple of Mackail's men ran after the fleeing man. A musket shot was fired and the man crumpled to the ground. Luke reached the fallen body just as Alistair arrived, his musket still smoking.

'He's dead,' announced Luke.

'Damn, I didn't want to kill him,' affirmed Alistair.

Luke turned over the body. It was Mackelvie. After further examination he spoke quietly to Alistair, 'You did not kill him. You musket shot only grazed his cheek. He has been stabbed. That is why he ran. He was escaping an assailant. Mackelvie was murdered.' Alistair rode along the line of people and isolated those in the vicinity of the incident. Another horseman, who Luke recognised as Duff Mackail, arrived. Alistair ordered him to move the remainder of the group away from the scene. Two servants affirmed that Mackelvie had been walking with Duncan Caddell on his left, and Mungo Macdonald on his right. Just before the killing a couple of men pushed through this group and temporarily broke up the trio. Just after the intrusion Aiden Mackelvie stumbled towards the shoreline of the loch. They did not notice any blood or unusual noise, but they were several yards behind the gentlefolk.

The suspects and witnesses rejoined the main group just as the ascent of a tall mountain began—and conditions worsened. A heavy snowfall escalated into a full-scale blizzard. On reaching an isolated shepherd's hut the women were placed inside, and the men found whatever shelter they could for the night against large rocks or in one or two small caves. It was a freezing. By mid morning the snowfall had ceased and the clouds lifted. Alistair called Luke to him and the two men carefully allowed their horses to pick their way to a large pinnacle that overlooked the path they had taken since the loch. Alistair unearthed a spyglass and was soon sucking his lips with interest. He passed the glass to Luke pointing to the down slopes of the adjacent mountain. Six horsemen were following them.

Luke immediately reacted, 'Who are they? They are not riding my horses, but by their gait they are definitely cavalry mounts. They must be David's.'

'Yes, they are probably your men on Scottish horses. Most of your horses were borrowed by Mackail and myself, as is the one you are riding. I will leave a few of Mackail's men here, who with the element of surprise and better position should be able to delay our visitors, or even force them to retreat. Once we reach our destination we will be safe from any number of horsemen.'

Several hours later Luke realised that Alistair had spoken the truth. The narrow path moved between a precarious edge and a precipitous wall. All of a sudden Alistair who had assumed the lead of the narrow column trotted over the edge. Luke's heart almost stopped. It was an illusion. Alistair had entered a tower house built into the side of the precipice. He entered across a drawbridge at the level of the tower's highest rampart, but some feet below the path. Once the party had entered, the drawbridge was withdrawn leaving a large gap of some twenty feet between the wall of the castle and the side of the cliff. In inclement weather if you were not aware of the castle's existence it would be invisible to any traveller on the path.

Eleven people sat around the table for dinner. Alistair and Elspeth took pride of place at the top and bottom of the table relegating Derek and Fenella to positions on their immediate left. Fenella sat next to Alistair, and Elspeth next to Derek. Luke sat on Elspeth's right. Luke observed the six remaining guests very carefully. Duncan and his protégé Janet appeared morose and did not initiate any conversation. On the other hand Malcolm and Mungo were voluble, anxious to express their disquiet at the situation. Young Morag remained oblivious to her new surroundings. The bright new face at the table was Duff Mackail who sat provocatively next to Fenella and at every opportunity pressed against her. Luke was jealous.

Alistair explained the situation. 'Forgive me for bringing you to Greytower. It will stop the Black Thistle communicating with the outside world. Castle Clarke leaked like a sieve and messages came in and out of the castle to almost everybody there. It was also important in Scotland's interests to reduce the influence of the English, who could at any time call up a shipload of troops to override our interests. However I have brought Colonel Tremayne and his sergeant with me to assist in the investigation, and to relay to the English authorities information that may concern them. Luke, I apologise for forcing you here, but you can understand my motives. Your investigation at Castle Clarke was going nowhere.'

Petrie angrily interjected, 'You tricked us, Sir Alistair. You told everybody that the English were about to send a ship laden with troops to Castle Clarke, and carry us off to Barbados.'

Luke replied, somewhat surprising himself by supporting Alistair, 'Sir Alistair may not have known the details but I can tell you that our fallback plan was for the English navy to return to Castle Clarke, and remove you all.'

It was Mungo who broached the question on the lips of them all. 'Sir Alistair, why did you shoot Mackelvie?'

Luke also had been troubled by Sir Alistair's quick resort to the musket. Alistair sidestepped the question, 'I did not shoot Aiden. He was stabbed while he walked. Duncan and yourself were the persons nearest to him, and therefore the prime suspects.'

'We were pushed aside by two or three men who stabbed Aiden, moved on into the crowd, and were not seen by anybody,' muttered Duncan, at last forced to comment.

'A very convenient lack of observation!' Luke said somewhat unfairly. 'Isn't there some thing any of you can report about what has happened?'

Fenella responded, 'Let's pool all the information, gossip or rumours we have regarding Aiden.'

Luke expected her ladyship to reveal much of the man's dirty past but she was stopped by Duncan, 'Death should be respected, my lady, and this is a matter rightfully reserved by our Lord to men. Women should not tax their weaker intellect with such concerns.'

Duncan's acolyte Janet joined the chorus, 'My lady, you and Lady Elspeth should join me in the reciting of psalms, and not waste your time on such worldly vanities. God will reveal Mr Mackelvie's murderer, as he will determine Scotland's fate. And he will consign to eternal damnation that cauldron of satanic, sectarian libertines, the army of the English Republic.'

The group looked furtively at Luke who was direct and brutal.

'Mistress Hudson, Mr Caddell has instructed you in the doctrine of Providence. God's will is revealed on earth in the fate of battles. The English army has won a series of amazing victories against all odds. Our most recent victory at Dunbar, where we crushed a vastly superior army, suggests to the neutral onlooker that God is on the side of the tolerant English and not that of the narrow minded bigoted Scots.'

Fenella and Elspeth both smiled at Luke's indirect attack on Duncan. Mungo however did not allow Luke's views to go unchallenged. 'Colonel, God did not give you victory at Dunbar. Scottish betrayal and Scottish incompetence was the cause.'

16

Malcolm Petrie was furious. 'You are not implying incompetence on behalf of the Scottish army or politicians? Dunbar was lost because the clergy overrode the military wisdom of the army commanders, and forced them to leave their dominant position on the heights, and descend into Cromwell's trap. Scotland lost Dunbar because of meddling clergy who should concentrate on bringing godly reform to the Highlands.'

'As I said, Scottish incompetence,' retorted Mungo.

Duff Mackail who had spent most of the evening pressing his leg against that of Lady Fenella quipped, 'This very discussion shows Colonel Tremayne why he will win, and subject Scotland once more to English rule. The lords of Scotland must unite behind the King, and reject the anti-royalist Kirk. A house divided cannot stand.'

Alistair nodded positively at Duff, while Derek who had taken no part in the discussion glared at his wife. She was being indiscreet enough to hold the Captain's hand as he outlined his views. Luke intervened, 'All of you want the King, but only on your terms, and you differ as to which faction should lead the new government under the King, and how much influence the King and the Kirk will be allowed. Who should rule? You all serve different masters—but which of you serves the Black Thistle?'

'For the moment it is more important to share what we know about Mackelvie,' said Alistair trying to refocus the discussion.

Derek made the initial contribution. 'Aiden was invited here by Duff and myself. We had unfinished business. He was a Campbell and owed his land to the expansion of that clan into Cameron areas over the last

89

century. His problem was that the powerful head of the clan, the Marquis of Argyle, was pursuing a policy opposed by Argyle's eldest son and heir, the Lord Lorne. Lorne wanted the Campbells and his father to throw their weight behind the King under the King's terms. Mackelvie was in turmoil. Surrounding Royalist clans were looking for the first opportunity to deprive the Campbells of their ill-gotten gains, which pushed Mackelvie towards Lorne. Yet at the same time real power still rested with Argyle, and if Mackelvie came out in support of the King then the Marquis might deprive him of his lands. His brother who was a right hand man of the Earl of Barr was sent into this area a month or so ago to negotiate with the Western Army. He may have added pressure on Aiden not to come out for the King.'

'Mackelvie was worse than a ditherer,' said Malcolm. 'He was a trimmer, if not a traitor. I have acted for the Scottish government in its dealings with the King. His agents on the continent listed Mackelvie amongst those who would assist the King on his return to Scotland. On the other hand the Scottish Government, including the Marquis of Argyle, considered Mackelvie was committed to them. Mackelvie was murdered because of his duplicity. The only question is whether it was on the orders of the King, or of Argyle and Barr.'

'So there are hidden agents of the King, and of Barr amongst us?' mocked Luke.

'Of course,' replied Alistair taking the bait. 'But we Royalists are not hidden. The English should not underestimate the loyalty and ability of the young King's supporters. We now dominate the Parliament, and have neutralised the Kirk. The King has dismissed many of the veteran senior officers. Young devoted Royalists now control the Scottish army that is camped at Stirling. There would be hundreds of men anxious to prove their new loyalty by dispatching a suspected enemy of Charles II. One such man may be in our company. I am loyal to the King, but it does not follow I killed Mackelvie, or endorse the plans of the Black Thistle to eliminate others.'

'But you did shoot at him without reasonable cause,' Luke emphasised.

Luke noticed a fleeting triumphal smile. It was on the face of Duff Mackail, or did Luke imagine it? Or as Andrew later commented, it could have more to do with his seduction of Lady Fenella, than his extermination of Mackelvie.

AT CASTLE CLARKE

David and Harry returned to Castle Clarke and were immediately informed of a large detachment of men moving along the loch's shoreline towards the castle. David ordered his men to assume a defensive position, ready to repel any assailants. He briefed Harry,

'They are armed Highlanders. They are certainly enemies of the English, and in recent times they have not been very supportive of the Scots army, although a few companies did fight at Dunbar.'

'Who are they?' asked Harry

'The array of blue berets suggests a well organized and financed operation with some links to our troops at Stirling. Blue berets were distributed to several regiments a few years ago including the Earl of Barr's. This is not a motley collection of hastily raised men for some sporadic raid against their neighbours. This is a military enterprise with a clear agenda which may be to dislodge us.'

'That is impossible given the defensive advantages of this castle, and its professional defenders.'

'Yes, that intrigues me. The Highlanders are not fools, and they would not commit themselves to a do-or-die attack unless it was a matter of clan honour.'

David took up a large spyglass and whistled as he watched the approaching army. 'Good news, Harry. The man on the chestnut stallion leading them is Sir James Cameron, chieftain of a cadet branch of the clan. I served with him over the years. But why are the Camerons marching on Castle Clarke which was garrisoned by Cameron supporters until our conquest of it?'

'Perhaps that's the answer. The Camerons are marching to avenge the eviction of their comrades,' Harry replied.

David kept watching the advancing Camerons and exclaimed, 'Sir James is confused. He is pointing to the flags we have flying, the Cross of St George, and that of St Andrew. To find the fort occupied by both Scottish and English troops has perplexed him.'

Sir James Cameron sent a horseman to the castle who demanded that Sir Derek Clarke surrender it forthwith. David replied, 'Inform your master that Sir Derek no longer occupies this castle, and that its current commandant

Major David Burns of the army of Scotland would welcome Sir James's presence to discuss the situation.'

Sir James obviously trusted David as he arrived without escort or companion and was soon drinking with the defending officers.

'What is going on here? A Scottish castle jointly occupied by English and Scottish troops. Where is that scoundrel, Sir Derek?'

David explained the joint enterprise and then asked Sir James, 'Why did you propose to attack the castle? To return it to Cameron control?'

'The Camerons have not controlled this castle for decades. I received an order from the Earl of Barr that the raiding that came out of this castle had gone beyond all acceptable bounds, and that Sir Derek and his men had been tried for murder and arson in their absence and found guilty. I am to carry out the court's verdict—execution.'

Harry struggled to speak, 'Sir James this is unwelcome news. When we captured the castle we were told the defenders were Cameron clansmen under Duff Mackail. We allowed them to withdraw. They subsequently returned and escorted the whole castle, including my commanding officer, into the Highlands.'

'These scum are not clansmen, but members of a vicious gang of cutthroats and blackmailers who raid the Lowlands to steal, vandalize and kill. Sir Derek nominally heads the gang but its effective leader is Duff Mackail, a renegade Cameron who was outlawed by the clan chief for his murder of clan rivals. Mackail is a killer several times over.'

'My god!' exclaimed David. 'We thought the inhabitants of this castle had freely left to escape possible imprisonment by the English, and were now safely ensconced in an isolated tower house protected by reliable and loyal Cameron clansmen. Now you tell us that the persons we were sent to rescue are in the hands of a bunch of cut throats.'

'It also clarifies the letter found in Oban. If that enterprise failed they could leave it to Duff to take care of Colonel Tremayne. Now Luke is in his hands,' conjectured Harry.

'I am more concerned for Sir Alistair and Lady Elspeth Stewart,' replied Sir James.

'You know them?' asked Harry.

'Yes, I am acting under orders from the Earl of Barr. He believes his daughter and son-in-law are under your protection. This is now more serious than you think. If Lady Elspeth's child is a male he will inherit the vast properties of his grandfather and the political power that goes with it. There are many persons, enemies and friends of the Earl who would not wish this child to be born—or if born, immediately placed under their physical control.'

'Sir James, do you know any of Sir Derek's guests—Aidan Mackelvie, Malcolm Petrie, Duncan Caddell, Mungo MacDonald, Janet Hudson or Morag Ritchie?'

'Three of the names are familiar, but I only know two personally. Aidan Mackelvie is the dominant local laird who represents Campbell interests in the area. He has seized Cameron lands, and is no friend of mine. I also know Sir Malcolm Petrie. He has for a decade represented the Scottish Government in its dealings with the English; the European powers; and the King, both father and son. There are rumours concerning his current loyalty.'

'Is he a Royalist stooge?' asked Harry.

'An offensive question young man. All Scots are Royalist. Sir Malcolm recently inherited a large amount of property from a second cousin. As with all Scots the defeat of the national army at Dunbar, and more recently the independent western army of the Kirk, have undoubtedly disconcerted him.'

'Could he be the Black Thistle's leader?' probed Luke.

'Maybe. When I was last at Stirling there were rumours that he had thrown in his lot with the King.'

'And the third prisoner that you know?'

'I have heard about young Morag Ritchie, although I have never met her. She is heiress to half a kingdom. At the moment there is a major conflict, legal and otherwise, to make her a ward of one of the senior government nobles before the nature of the administration changes. The Earl of Barr currently administers her lands.'

'What do we do now?' David asked, as much to himself as to his companions.

'Rescue the prisoners! They are victims of a vicious gang who have no conscience or honour. The Black Thistle moved them to Greytower where he can kill them all,' exaggerated Harry.

'Why would he want to do that? This man has a political agenda not a personal vendetta,' David commented with surprising intensity.

Harry responded firmly, 'I am not so sure. From the little we know this could be a personal vendetta against the Earl of Barr. The Black Thistle has the man's daughter, son-in-law, and wealthiest ward.'

David continued to argue passionately for a political explanation, 'No, it is an attempt to dissuade the Earl from moving his position anywhere except towards the Royalists, the only option for Scottish nationalists in the current situation. You can blame the English for this dilemma. Until the Battle of Dunbar, the Kirk dominated government did not need young Charles. Then with the creation of the Western army of fanatical Presbyterians his assistance did not appear so vital. This situation was destroyed by General Lambert when he annihilated the western army a month or so ago. The Black Thistle is using the situation to apply further pressure on the leading nobles to welcome the King and his Royalist followers into the army and the government. In reality you are witnessing a Royal coup. The situation remains political and not personal.'

Sir James intervened, 'In either case it is a Scottish problem. Why are the English involved?'

'Somebody was sent from this castle to warn the Scottish government that there was a plot afoot to remove not only leading Scottish politicians but also the English high command. The Scottish government knew it could not capture Castle Clarke without the help of a powerful English warship. Our political concern is to prevent the assassination of our generals, and now it has become personal with the abduction of our commanding officer, Colonel Luke Tremayne. The Black Thistle outmanoeuvred us by moving into the mountains,' admitted Harry.

'We must move on the mountain retreat without delay. I act in the judicial interests of the current government as must you Major Burns. And the English want to rescue their colonel.' Sir James concluded.

'Gentlemen, technically this castle is in English hands. We expect *The Providence* to return tomorrow or the following day with supplies and more troops. David, your men could go with Sir James. I will follow with Colonel Tremayne's personal troop in a few days. As you rightly imply this is a Scots enterprise, and our English concern is limited to the rescue of our two officers.'

The Highlanders appreciated the warmth and hospitality of the castle, but at first light, which was well into the mid winter morning, the combined

Scottish units of Cameron Highlanders and Earl of Barr Borderers left for Greytower. While Harry Lloyd was forced to wait for the arrival of more troops and supplies, he nevertheless sent Sergeant Halliwell into the mountains with orders to reach Greytower before the Scots.

17

Greytower

Luke and Andrew shared a tiny room, and were awakened by Sir Derek hammering on the door. 'Is there a problem?' asked a serious Luke.

'Her ladyship thinks so,' announced Derek almost apologetically. 'We gathered for breakfast at the usual time. Fenella was cross that so many of our number were missing. She sent a servant to ascertain the situation with Sir Alistair and Lady Elspeth. He reported back that they were missing and their beds had not been slept in.'

The soldiers were soon in the dining hall where the other residents were slowly taking in the news that the Stewarts had disappeared. When Alistair assumed a dominant role within Greytower Luke had imperceptibly become his partner in attempting to solve the problems surrounding them. Duff Mackail now usurped control, and made it clear that the English were prisoners, and not partners. He ordered them to join the rest of the residents confined to the great hall.

Before Luke could protest Duff disappeared with his men to search the castle, leaving one callow youth to guard the residents. Sir Derek was less antagonistic and allowed Luke and Andrew to join the search. As soon as they left the hall they headed for the lower levels of the tower which had not been inhabited in recent times. Luke held up his hand and placed his finger to his lips. Ahead of them two soldiers were speaking in an agitated manner. Luke took the initiative. He made a deal of noise and bounded down the spiral staircase to confront two of Mackail's men. They were towering over the prone body of Elspeth. Luke sent the men to inform Mackail. Luke placed the blade of his dagger across Elspeth's mouth and smiled, 'She is alive.'

The two Englishmen carried Elspeth up the steep staircase to her bedchamber where Fenella, alerted by Mackail's men, was waiting to minister to her. Fenella asked the men to leave the chamber while she undressed the unconscious woman. Luke refused. 'Fenella, we will avert our eyes, but I am not leaving here until she regains consciousness, and explains what happened to her. If this was an attempted murder the would-be killers might strike again.'

'Don't exaggerate Luke! Elspeth probably slipped on the derelict staircase. One of my servants, as you know, is a cunning woman. I have sent for her to assist me.'

The cunning woman arrived and removed an array of herbs and ointments from her basket. These she rubbed gently on Elspeth's obvious bruises and wounds. She then mixed a potion in a small cup, and put it to the lips of the victim. Luke suddenly struck the cup from her hand and sent it flying across the room. The cunning woman was taken aback, but Luke motioned to her to continue her administration of external potions, but ordered her not to give Elspeth anything internally. Several hours later Elspeth began to stir and demanded to see her husband, but soon relapsed into a deep sleep that worried Luke. Military surgeons did not allow soldiers with head injuries to fall asleep, and Elspeth may have suffered head injuries. Should he wake her up? He let her sleep.

After asking Andrew to take over his vigil at Elspeth's bedside, Luke joined the rest of the castle for the evening meal. Everybody wanted news of Elspeth's progress. Luke was honest. 'Lady Elspeth recovered consciousness, but has now fallen into a deep sleep. When she awakes she will either be her old self, or very, very ill.'

'Why wander around the lower levels in the middle of the night? No wonder she slipped!' proclaimed an unsympathetic Simon Caddell.

'Elspeth may have been tricked into going to the deserted lower levels, and then pushed further down the staircase. This is another attempt by the Black Thistle to kill its enemies,' announced Luke, hoping to provoke a meaningful response.

It did. Mungo asked, 'Colonel, you surely do not believe that one of us tried to kill Lady Elspeth?'

'Who else is there? responded Luke, who then turned to Duff and asked, 'Any luck in the search for Alistair?'

'He is not in the castle,' replied Duff without emotion.

Later that evening Luke resumed his position at Elspeth's bedside. His nightly vigil was eased by the provision of a whisky based honey flavoured drink and some sweet oatmeal cakes. Just after dawn Elspeth awoke. Luke was delighted. She was alert, and while claiming a headache she did not reveal any injuries other than the bruises she suffered in her fall. She called for female help to assess whether her child still lived. Luke acquiesced but before they arrived he placed his head against Elspeth's and whispered, 'Forgive this intimacy my lady but your life is in danger. I do not trust any of your companions.'

'Get Alistair! He will protect me.'

Luke explained that Alistair had disappeared at the same time as she suffered her fall. She began to sob and then quietly confessed, 'I fear the worst. Yesterday Alistair overheard two of Duff's men talking, and came to me very agitated. He felt that the move to this castle had been a mistake and he been badly misled by everybody. He confessed that the only people he could trust were his enemies, the English.'

'Did he tell you what he overheard, or why it led him to distrust the rest of the company?' asked Luke.

'No, but he decided to explore the castle to see if there was a way to escape if necessary. Just before we retired for the night he went off to investigate. When he had not returned after some time I went in search of him. I took one of the tapers from the hall and after I had descended a level I heard a noise behind me. I continued down to the next flight where a sudden gust of wind extinguished my taper, and in the dark something crashed heavily into my back. I don't remember anything else until I awoke here.'

'Why try to kill you?' asked Luke.

'If they wanted to kill me they could have made sure of it by using their swords and daggers.'

'Then what is this about?'

'The baby,' Elspeth replied with a tear in her eye.

'You were assaulted to abort the birth of your expected child?'

'Yes. My father, the Earl of Barr has no son. If my child is a male he will become the heir to the most extensive and wealthy estates in Scotland and in time the immense political power that would bring. The existence of my son could change the history of Scotland,' announced Elspeth with a modicum of pride.

'Is your abduction and this subsequent attack personal, to punish your father, and protect the interests of potential heirs: or is it part of the political campaign to ensure that his power is not continued into future generations?'

'I cannot understand a political attack. Father has remained within the mainstream of Scottish nobles in support of the Scottish parliament for the last decade. He has supported the coronation of Charles II as King of Scotland but under the tight restrictions of the Kirk led government. He is neither an extreme Royalist, nor an extreme supporter of the Kirk. I cannot see what political motive the Black Thistle has in punishing him through me, especially now that his political power is on the wane. This is personal. His rise to power and the complex relationships have always sparked jealously, feuds and conflict within the larger family.'

'Who is currently heir to your father's lands?'

'Alistair will inherit most of it on the death of my father. However some of father's land was obtained under different dispensations that would give five or six other family members substantial holdings. However if I give birth to a male, all of father's estates will eventually go to my son. Even though Alistair should legally inherit the major part of father's estates there are family members, particular holders of land on the borders, who would resent it falling into the hands of a man who is not a blood relative of the Earl. They may fight to prevent Alistair, a Stewart, taking over.'

'What if Alistair dies and you have no child?' asked Luke.

'As a widow I become the heiress to the whole estate once again until such time as I might remarry.'

'And then your new husband and his family will inherit your father's wealth and power.'

'Yes.'

'Could your own father, realising his mistake in marrying you to a Stewart, want Alistair killed, so that he could control his ultimate heir?'

Suddenly Duff Mackail burst through the door, 'I am sorry to interrupt you Lady Elspeth. Colonel, my men have apprehended an English soldier who claims he brings a message for us, but who refuses to reveal it without your presence.'

On reaching the great hall Luke was surprised to see John Halliwell. As they hugged John whispered, 'Don't believe a word of what I say.'

It was Duff who broke the moment, 'So sergeant! What's your message?'

'Captain I am not sure to whom I should direct my information. My acting commander Lieutenant Lloyd asked me to speak to Sir Alistair Stewart, I understand that he has disappeared.'

Sir Derek intervened, 'I am head of this household. What is your message?'

'Lieutenant Lloyd wishes to inform you that Major Burns and his men have joined a large company of your clansmen under Sir James Cameron and are heading here to assist you.'

Luke picked up on the shocked expression on the faces of both Duff and Derek. The latter quickly recovered and asked, 'Why would an English lieutenant send you all this way to tell us such an unimportant piece of news?'

'You are very perceptive Sir Derek. Lieutenant Lloyd was concerned about Colonel Tremayne and Sergeant Ford. He hoped that I would be able to return to Castle Clarke with them, or at least take a message back that they are well.'

Derek burst into a big smile, 'An excellent suggestion soldier. It was Sir Alistair who brought the Englishmen here. I much prefer to keep things solely in the hands of the Scots. All three of you can leave in the morning.'

Duff half raised his hands in protest, but thought better of confronting Derek in front of the Englishmen. Luke and John took their leave and returned to Luke's chamber. Once inside Luke asked John why Harry had sent him on such a dangerous mission. 'It is even more dangerous than you think,' replied John.

'Why should news that Duff is being reinforced by more clansmen be a problem for us?'

'Because Duff and his men are not clansmen. They are brigands that prey on stock and properties throughout Western Scotland. Sir James Cameron is coming here to hang Derek and Duff on the spot.'

'Then why did Harry warn these villains?' asked Andrew.

'He hoped Derek would react the way he did. Reduce the number of his enemies within the castle by allowing you to leave,' John replied.

'But we cannot leave,' Luke relied sternly. 'Sir Alistair has disappeared, feared murdered, and Lady Elspeth has been attacked, and is in constant danger. I will not leave without Elspeth. I doubt that the Black Thistle, who increasingly looks like the jovial rogue Derek, could really allow it.'

The three Englishmen decided to protect Elspeth. They persuaded Janet Hudson to sit by her side. Luke took up a position inside the chamber beside the door, while John and Andrew took it in turns to stand guard outside the chamber door. While not on duty the other sergeant slept on a mat that had been placed across the entrance to the room. Luke remained anxious. The three of them were prisoners of two ruthless men who were supported by a dozen henchmen. Luke hoped to change the balance a little next morning by bringing Duncan, Mungo and Malcolm into their confidence.

Luke arrived at breakfast ready to implement his plan.

18

Elspeth, now feeling well enough to dress, and a tired Janet had made their way to breakfast where they joined Fenella, Malcolm and Mungo. Luke sensed there was a problem. Fenella was highly elated.

'Welcome Luke. As the highest-ranking officer remaining here, you are now in command. Derek, Duff and his men have left. Derek told me that as Sir James Cameron would soon be here, Duff and his men were no longer needed to watch the prisoners.'

'And where are Simon Caddell and Morag Ritchie?'

'I assume they left with Derek,' replied Fenella.

'Willingly or forced? Lady Elspeth and Mistress Hudson are still here because my men protected them throughout the night. I fear that Caddell and Ritchie have been abducted for a second time, and Sir Alistair has disappeared. However Sir Malcolm and Mungo Macdonald are still with us.' Luke turned towards them and asked,

'Gentlemen why are you still here and Caddell and Ritchie gone?'

'Caddell and young Ritchie may have asked to go with Sir Derek. The thought of wild Cameron Highlanders may have frightened them. As far as I know they have no friendly links with the Camerons,' replied Macdonald.

'Neither of you were approached by Sir Derek or Mackail during the evening?'

Malcolm was blunt. 'They could not reach me if they had tried. My chamber has the only working bar lock. I placed the thick wooden bar into the solid metal clasps. It would take a battering ram to dislodge that door.'

'What about you Macdonald?'

Mungo was agitated, but having settled some internal conflict decided to tell all. 'Last night Sir Derek and Captain Mackail entered the great hall and sat at this table. They did not realize that I was in the chamber, hidden by the back of a large chair into which I had slumped. They were greatly alarmed, to the point of panic, by the news that Sir James and Major Burns were heading here with over a hundred soldiers. Their dozen men could not hold out against such odds, especially as there were three heavily armed Englishmen within the castle. Sir Derek and the Captain are apparently not whom they pretend to be. They were convinced that Sir James was coming to hang them for their raiding activities. I was petrified when Captain Mackail suggested that they kill all the prisoners before they left. Derek was sarcastic, suggesting that the last attempt had failed, and they did not have time. Above all the English protected those that should be killed. They agreed to take as hostages everyone they could find. As soon as I heard them leave the chamber I disappeared at full speed into the depths of the tower. I did not come back until daylight.'

Andrew reported that the relieving troops were making their way through a narrow pass and would soon be at the castle. Luke rode out to meet them. Sir James was furious that he had missed the criminal gang. Luke lied, indicating that they were alerted the previous afternoon by the sight of the clansmen on the adjacent mountain. Luke, David and Sir James discussed their next move. It was imperative to follow the gang and rescue Caddell and Ritchie. David assumed control of Greytower and its inhabitants. The Englishmen would accompany Sir James in pursuit of Derek and his gang.

Before he left Luke spoke to Elspeth who, according to the women of the household, had not lost her baby through the fall. He immediately reported to David and Sir James, 'Elspeth must be escorted to her father as soon as possible. Clearly she is a victim, and the murderer may strike again. As we cannot be sure of Janet, Mungo and Malcolm they should be kept here for their own security, and to prevent them having any access to the outside world until the situation clarifies.'

Sir James agreed, 'Keep them and Lady Fenella under constant surveillance. I will leave most of my men behind to give you additional manpower. Derek has headed further into the mountains with only a dozen men. I only need the same number to deal with him.'

'Before we leave Sir James, could you have your men and those of Major Burns search every inch of Greytower? Alistair may be hidden somewhere in its depths. There are parts of this castle that have not been used for decades,' requested Luke.

There were reports of secret chambers, strange tunnels and unexpected exits as over a hundred persons search every inch of the edifice. Two hours after the search began one of David's men re- entered the dining hall where the officers and castle guests remained soaking up the warmth of a blazing fire. Although the man had come up from the depths of the castle his cloak revealed numerous thick snowflakes. He whispered to David who immediately signalled to Sir James and Luke to follow him.

They descended the staircases to a level below that where Luke and Andrew had found Elspeth. They then reached a blank wall. The soldier pulled on a stone head carved into the wall, and the wall swung outwards allowing swirls of snow to enter the castle. This hidden door opened on to the ledge on which the tower had been constructed. Visible against the precipice. and at the end of the ledge was a large cave. The group entered the cave. Lying just beyond the entrance was the prone body of man whose head lay in a pool of blood. His throat had been cut. Luke felt a pang of sympathy for Elspeth. David turned the body over. There was a gasp from Luke. It was not Alistair. It was Simon Caddell.

Luke eventually climbed back up the stairs and informed Elspeth. She burst into tears and her whole body progressively relaxed with relief that the body was not her husband. Luke asked, 'Did Caddell have any link with your father?'

'Not that I know,' replied Elspeth.

'David will escort you to your father as soon as possible. I am about to leave with Sir James in pursuit of Derek and Duff.'

'I will not go, Luke. I will stay here until Alistair returns,' declared a defiant Elspeth.

'Beware of Fenella. She lied to us about Simon leaving with her husband,' warned Luke.

'Maybe she was lied to, and assumed that as Simon was missing he must have gone with Derek. Is he the Black Thistle?'

'He and Duff Mackail are certainly members.'

'You are wrong Luke. Derek and Duff are simply criminals. They are out to make money. They have no reason to kill Aiden or Simon on their own account. They were paid to do this by someone with money, or by the Black Thistle. Its agents are still here and that frightens me,' admitted Elspeth.

'I will alert David to your fears. Be careful!' Luke gave her a brotherly hug and departed.

Sir James and the three Englishmen rode into the blizzard, with twelve heavily armed clansmen jogging behind them. An alert Sir James was immediately concerned. 'Tremayne, these are the tracks of five or six men at the most, half the number you led me to expect.'

'Clarke's party must have split somewhere along the path,' said a puzzled Luke.

'We have just come along it. There was no obvious place for a group to leave the main party although in this snow visibility is reduced to a few yards,' Sir James replied.

'Then they must still be in the tower house,' surmised Andrew.

'No. We searched it thoroughly. They must have hidden along the path as we approached the castle, and then after we had passed they headed back towards the coast,' announced Sir James emphatically.

'In which group are Derek and Duff?' pondered Luke aloud. Sir James thought for a while, 'We must consider both groups as important as each other. Colonel, send one of your sergeants back to Greytower and ask Major Burn to send a party of my men downhill after the fleeing brigands.' Luke nodded to John who pulled his horse tightly against the cliff face as the clansmen moved past him. He turned his horse and headed back towards Greytower.

The snow stopped falling, the clouds lifted and the weak sun lit up the face of the mountain opposite. Luke saw a party of four beginning to climb it. One large man was on a horse and the others walked. Sir James trotted down the slope of their mountain at a pace which forced the clansmen into to run. 'We will catch them by early next morning. Derek's men are not used to raw Highland conditions.'

Andrew sympathized with the fleeing criminals. He had not faced such cold since he left Sweden a decade ago. Luke had never experienced such conditions and despite his adequate clothing his hands were numb. He was concerned as he observed the fleeing brigands. None of the figures opposite appeared to be that of a small girl. Morag must be with the other party. Sir James was having similar thoughts. 'The girl does not appear to be with this group, although the figure on the horse is either wearing a large cape, or he may have the girl mounted in front of him. We are still too far away to tell.'

As darkness fell Luke expected Sir James to stop for the night. He did not. As light faded totally he turned to Luke. 'My horse is surefooted and can find its way unerringly along narrow paths such as this. Put your horse right behind mine and have your sergeant do the same with yours. My men know how to move at night, but let us get off this mountain as soon as we can. The extra hour or so gained now will make tomorrow so much easier.'

Luke was unhappy. He felt once more that loss of control he had experienced deep in the American woodlands. The unknown was overpowering. Edging down a steep and narrow mountain track, with the snow once more beating into his face was frightening enough. But to do it in pitch darkness, relying solely on the ability of his horse to mimic the animal in front was totally unnerving. Eventually they made camp for the night and Sir James offered Luke a swig of a flavoured whisky. Next morning before the sun had risen they were again on the move. The snow stopped, and the clouds lifted slightly. Luke could see their quarry not far ahead of them. Then it struck him. The group ahead was seriously diminished. There was no horseman. It was mid morning when Luke and Sir James rounded a bend on the narrow upward path. They were confronted by four stragglers who immediately raised their hands in surrender. The cold had got to them, as they apparently were not able to light a fire. They were half frozen. They had left Greytower in such a hurry to escape the fearsome Cameron Highlanders that they had not enough time to adequately clothe and equip themselves for the extreme conditions they confronted.

Luke turned to Sir James, 'Before you send these men back to Greytower let me question them about their missing companions!' Luke grabbed one of the shivering brigands and with one hand around his throat and a dagger pressed against his heart asked simply, 'Where is Sir Derek, Captain Mackail and the girl Morag Ritchie?'

The traumatised man simply gestured along the path. 'Our leader and the girl have pushed on, but on foot. The horse died during the night.'

The brigands were not returned to Greytower. They were lined up and simultaneously decapitated. Sir James ordered his men to return to Greytower while he, Luke and Andrew continued the hunt. Within twenty minutes they came upon a heavily hooded caped man and a young girl making slow progress. They were thigh deep in soft snow and sank further as the soldiers watched. The girl turned and cried out for help. Luke moved in for the rescue. Sir James grabbed his reins. "You'll die if you ride your heavy cavalry steed into that corrie. It is a large deep hollow filled with soft snow.'

The sinking man turned and faced them. He placed his sword on the snow and threw back his hood revealing a swarthy bearded face. It was not Derek, nor Duff. Andrew recognised him as one of Duff's deputies. Even more disappointing was the identity of the young woman. She was not Morag but a maidservant from Castle Clarke. Derek and Duff had won. Their diversionary ploy had worked. The brigand leaders had escaped with their prisoner, and days had been wasted on a futile pursuit. Sir James dismounted, primed his pistol, walked gently across the soft snow and methodically shot the swarthy one between the eyes.

19

He was about to repeat the exercise with the young woman when Luke intervened, 'Sir James desist! She may be able to give us information.' The girl clung to Luke as he dragged her from the snow. He carried her out of the corrie and placed her in front of him astride his horse. This intimate situation gave the young woman security, and Luke an intimacy to conduct a subtle and gentle interrogation as they moved slowly back to Greytower.

'Have you known Lady Fenella long?' probed Luke.

'No. I have been in service since I was eleven but two years ago my master transferred me to Mistress Fenella on her marriage to Sir Derek, and her forced relocation to the edge of the Highlands.'

'What did you know about Fenella before you went into service with her?'

'The common gossip was that she had been disappointed in love, and that her marriage to Sir Derek was very suddenly arranged.'

'Did your old master tell you anything about her?'

'Yes, she had been brought up on the estate of a distant cousin to whom my master was related and whose property was not far from the Border. Her mother had died in childbirth and her father, a busy government lawyer and official, had no time or estate to nurture her. When she became a young woman she moved to the house of a notable in Edinburgh to find a husband.'

'What's this gossip about disappointment in love?'

'Given her less than affluent background an excellent marriage was negotiated for her with some distant, but wealthy relative of the Marquis of

Argyle. At the last moment the potential husband married someone else and Fenella was left distraught, and in that state was married off hastily by her embarrassed relatives to that rascal Sir Derek.' Luke's questioning was ended when Sir James ordered them all to dismount and take shelter for the night. They had reached a ruined building whose sole standing wall protected them from the bitter wind and flurries of snow. A fire enabled the men to heat up a pot of snow to which oats and chunks of dried meat were added which in time created a thick potage that if nothing else was warming.

Luke observed the young servant who identified herself as Gillian Shaw. Although she was still hooded and heavily cloaked Luke found her a very attractive woman, and hours of pressing his body against hers had physically stimulated him. Luke pushed back her hood. Gillian had a snub nose, large inviting lips which despite the cold were naturally red, a wide face, hazel eyes and flowing chestnut hair that now cascaded down to her shoulders. He was determined to see a lot more of this woman, but the current situation was neither appropriate nor politic. He had no desire to alienate the girl who might still hold useful information.

The group settled down for the night, utilising all available body heat to survive the conditions. Luke made his body available to the petite Gillian Shaw who did not object to Luke encompassing her in his broad arms. As the night progressed Luke found his hands caressing what he discovered were firm taut breasts and this in turn led him to react. This reaction intensified as her hands began to caress his aroused genital. It did not take long for the combination of her hardening nipples and intense manual stimulation for Luke to explode. Unromantically within minutes he was sound asleep.

Next morning they continued their journey. Gillian besotted with Luke pressed back into his body as soon as they were mounted. While enjoying the stimulation Luke resumed his questioning. 'Since the prisoners and guests arrived at Castle Clarke some months ago have you seen Lady Fenella show any sort of favouritism or special treatment to any one of them?'

'Where do I start? Lady Fenella had encounters with most of them.'

Although Luke could hardly conceal his eagerness for more information the snowfall had changed into a blizzard and it was impossible to hear what Gillian was saying. They settled for the pleasantness of physical contact. Late on that second day they reached Greytower. Luke indicated he would call for Gillian after they had eaten. David and John listened to the report that Sir James and Luke delivered over the evening meal. The soldiers ate separately from the dwindling number of civilians. Only Lady Elspeth, Lady Fenella, Janet Hudson, Sir Malcolm Petrie and Mungo Macdonald remained. Of the original inhabitants of Castle Clarke, Sir Derek and Duff Mackail had fled taking Morag Ritchie with them; Duncan Caddell and Aiden Mackelvie were dead, and Sir Alistair was missing. It was a messy situation.

David explained that the Highlanders led by one of their junior officers had moved off on their mission to track down the other half of Derek and Duff's criminal gang. Sir James with the remainder of his men would catch up with his comrades and complete his judicial task. David's unit, reduced to one company would remain responsible for security in Greytower. After the capture of Castle Clarke the remaining Scots regiments had returned to Stirling. The three English officers would continue investigating the murders, and attempt to uncover the identity, and plan of the Black Thistle.

When the meal was finished the Scots officers left to organize their troops. Luke discussed with John and Andrew the potential value of Gillian Shaw's evidence. 'And I thought your interest in her was purely physical,' quipped Andrew who winked knowingly at John. 'And what has she told you of value?' asked John.

Luke openly admitted, 'Nothing much, except some details of Lady Fenella's past.' He went on to explain that, 'She was about to tell me more but it was difficult to continue our discussion on horseback during a blizzard. I will continue my interrogation this evening.'

'Would you like us to take over the questioning?' said a grinning John.

Luke did not answer, and a chuckling Andrew led John away to drink in front of the roaring fire on the upper level. Luke found Gillian who had just completed her chores. The young woman beamed when she saw Luke and immediately followed him back to the chamber he had just left. Luke motioned for her to be seated. He remained standing and asked, 'Gillian you were about to tell me about Lady Fenella and her relationship with the inhabitants of Castle Clarke. How did she react to the two murder victims, Mackelvie and Caddell'

'Surprising that you should raise them together because they were the only persons, apart from Sir Derek that I overheard Lady Fenella argue with.'

'Continue.'

'I approached my mistress's chamber one evening and heard her shouting, "How dare you humiliate me a second time." A voice, high pitched and squeaky which I thought was a woman, but now recognise as Mackelvie protested, " I am not here by choice." She shouted at him to leave the castle as soon as he was able, and in the meantime avoid her presence as much as possible.'

'Did you hear any further altercations between them?' asked Luke.

'Not exactly, but on several occasions she was very rude to him which led several of the other guests to raise their eyebrows.'

'What about Caddell?'

'She told me, most unusual for a mistress to comment on a guest to a servant, that Caddell was a sanctimonious hypocrite. He had made many enemies and would come to a sticky end. She was determined that he would not dominate poor Mistress Hudson.'

'She had the opposite relationship with Duff Mackail?' probed Luke.

Gillian blushed and answered more softly than before, 'She and Captain Duff are lovers. I knew of their regular meetings because each time Duff visited her ladyship his deputy whom that Highland officer shot on the mountains, came to me.'

'Yes Gillian your own behaviour needs explaining. Why did you run off with your friend, when he left here in such a hurry? Were you so much in love with him?'

'What's love? He looked after me and after their raids he always brought me presents. I ran away with him because he told me that an army of wild Highlanders was marching on the tower to wipe out all Lowland Scots and Englishmen. To save my life I must go with him.'

'Did Lady Fenella show any special reaction to Mungo Macdonald or Sir Malcolm?'

'She kept her distance from Macdonald. She might have been frightened of him.'

'And Sir Malcolm?'

'That was the greatest surprise. I found her in a tender embrace with Sir Malcolm. He was stroking her hair and she was hugging him and crying. They did not break apart on my arrival, and ignored my presence. No words were being spoken.'

'Was this a sensual sexual embrace, or that of a kindly fatherly figure to a distraught younger woman?'

'How could I tell?' replied Gillian.

'The two men that Lady Fenella disliked are dead and that the two she seems to have confided in are still with us. Did she have Mackelvie and Caddell killed?'

'She had the means and opportunity. Sir Derek, Captain Duff or any of his men would act for her.'

'Did your swarthy friend ever mention orders to remove any of the inhabitants of Castle Clarke?'

Gillian blushed again and lowered her eyes. 'He mentioned once that their orders to murder the Englishmen, especially you, were on hold because of the interference of Sir Alistair.'

Luke especially noted this comment and continued with his questioning, 'How did her ladyship relate to the women?'

'She was very jealous of Lady Elspeth. She feared that Lady Elspeth would attract Captain Duff, and be more popular with the men than herself. Also she resented Lady Elspeth's superior status as a daughter of an earl. Lady Elspeth's presence in the Castle disconcerted Lady Fenella. On the other hand she showed nothing but kindness for Mistress Hudson, whom as I said earlier she was anxious to protect from Caddell, and for young Morag Ritchie whom she treated as a daughter.'

Luke moved around the table and gently raised Gillian from her chair and held her closely to him as she responded with the most passionate of kissing. Both parties were lost in the moment which was suddenly shattered. Elspeth entered the room leaving Gillian to scurry off to her quarters. Her ladyship commented sharply, 'The mistress must be disappointing if you have to resort to the servant.' With that she left the room before Luke could reply, although her broad smile was inviting rather than censorious. Luke retired for the evening. He now shared a small chamber of the tower house with David, Andrew and John where they slept on the floor, on top of thick animal skins and with no shortage of blankets. David was taking his turn on guard duty and both John and Andrew were snoring, but not quite in unison. Luke too was soon asleep.

Sometime during the night he stirred. He sensed a presence towering above him and reached for his dagger. Before he acted he glimpsed a female form in a long chemise and wearing a hood and cape. She grabbed his hand and led him out of the room. He was led down the stairs to the bedchamber of one of the guests. He followed the mysterious woman into a room where a lone candle flickered dangerously. As he entered the room, and was led to the bed the woman threw off her cape and hood and pulled Luke after her onto a large four-poster bed. To Luke's surprise the woman was not Gillian.

20

t was Fenella. Twenty minutes later both fell back exhausted. Luke remarked, 'That my lady was unexpected. What do you want?'

'Luke, you can completely destroy a moment. With the desertion of my husband and lover I am in desperate need of your company.' Fenella was irresistible. Luke allowed himself to embark on another rapidly escalating episode of lovemaking. Eventually a cacophony of gentle snoring indicated that both were fast asleep. It was still dark when they awoke. Fenella whispered gently in his ear, 'Will you hunt down Derek and Duff?'

'Finding Derek and Duff is a Scots affair, and Sir James intends to execute them on the spot. It is only a matter of time. What you do in these circumstances is a more pertinent question.'

'I will go to my father's new estate,' was the unemotional response.

'Although I am not interested in Derek or Duff, I am determined to find young Morag. Can you help?'

'And why would I?'

'Because of your kindness towards the young girl. Accompanying Duff and your husband through horrific terrain, being chased by Sir James does not augur well for her survival.'

'I agree, but to be fair to Derek he did not want to take her. Duff disagreed.'

'What did Duff want?'

'He wanted to keep the girl with them as a hostage, and sadly, if she became a burden, he would kill her. I sided with my husband by suggesting another solution.'

'Which was?'

'Talking won't help. I will take you to her.'

'Is this some ploy for you to escape Greytower, and be rescued by Derek or Duff?'

'You are a cynical curmudgeon. It's up to you. If you want to rescue Morag then trust me. Tell the others you are escorting me into a neighbouring glen to visit a sick relative.'

'You and I travelling alone in blizzard conditions in these isolated mountains is not safe. I will bring my sergeants with me, and we must take plentiful supplies.'

'And I will bring a servant to help me minister to my sick relative.'

Within an hour Luke, Andrew and John with Lady Fenella and her servant, Gillian Shaw, made their way down the mountain, across a small glen and then began the ascent of a large mountain. Fenella explained that their destination was the large glen that could be reached the following day. Luke recognised the glen as the one in which Mackelvie had been murdered. It now dawned on Luke that his assassins might have been locals that had been paid by an inhabitant of Castle Clarke.

Luke was anxious. They were being watched. What a fool he had been! How could three Englishmen defend themselves against dozens of locals? His temporary lust for Fenella had warped his judgement. On reaching the large glen Fenella led them off the main road. Suddenly there emerged from the trees a dozen armed men led by an impressive looking woman. Fenella dismounted and engaged in a long conversation with a tall sprightly younger woman whose plaid cape and matching dress was a colourful combination of bright yellow and vibrant blue. What was this woman of apparently high status doing in a poverty stricken hamlet? Fenella motioned Luke and his men to follow her. The younger woman led the visitors to a house that was slightly bigger than the normal peasant hut. Fenella introduced them, 'Colonel Tremayne, this is Donna Mackail, Duff's sister.'

The women removed their outer garments, and a servant brought hot mugs of a thin beef broth. She turned to Luke. 'Another of Fenella's conquests. No sooner is my brother gone than she finds another man, and an Englishman to boot. Fenella have you no shame?'

'Come Donna, you will give the Colonel the wrong idea. He is not deep in the Presbyterian lowlands. Highland women have always been generous, but enough of this banter. I am here for Morag. She is a fragile little girl, and the murder attempt did nothing to improve her condition.'

'Why would I hand her over—even if I had her?' said Donna with a mischievous smile.

'Because she is another mouth to feed, and her death would not endear you to Derek.'

'But it might to Duff, who hinted that her death would be no great loss.' Donna then asked Fenella and Luke to follow her through the house to the adjoining barn. Morag was lying on a straw palliasse, and covered in blankets. This was the warmest room of the dwelling, although the stench from the animals that shared the chamber was overwhelming.

'Is she well enough to travel?' asked Luke concerned by the pale and motionless body that he observed.

'How far would you take her?' asked Donna.

'Castle Clarke and Greytower are about the same distance away, but conditions are better at the Castle,' Luke noted.

'No Luke. She returns to Greytower with me, unless you are happy for me also to go back to the Castle,' interjected Fenella. Morag eventually joined the others in the entry chamber and after consuming a bowl of hot soup immediately looked better. Gillian took charge of her and was in deep discussion concerning her needs for a journey when the door to the barn burst open. A man, with his musket primed, burst through the entry.

'Well done my precious!' Sir Derek gave his wife a peck on the cheek, and Donna removed Luke's sword and dagger while a couple of locals disarmed John and Andrew.

'I have been well tricked my lady,' remarked Luke with guilt about his own stupidity, rather than fury with the conniving seductress.

'Be quiet Luke. You completely misunderstand the situation. You and I will return to the Greytower with young Morag. There will be a slight delay while you negotiate with Sir Derek, whose request you will find easy to comply with.'

'So you inveigled me here to negotiate with Sir Derek?'

'Only in part, I too want Morag rescued and restored to health.'

'Well, what do you want Sir Derek?'

'Very simple Colonel. My men have deserted to Duff. I want you to arrest me, and take me to Castle Clarke which I believe is entirely in the hands of the English. I want to be a prisoner in my own castle protected by the cream of Cromwell's army. That is better than being butchered by Cameron's highlanders somewhere in this mountain wilderness. Under Scottish law I am a dead man.'

Luke acquiesced. To have Fenella and Morag safely back at Greytower, and Sir Derek a prisoner of the English were both positives. 'My party, with the exception of Sergeant Halliwell, will return to Greytower with Morag. John Halliwell will escort you with some of the locals to Castle Clarke. I will tell Sir James that you surrendered to the English in the act of voluntarily handing over Morag to us. Halliwell will put Lieutenant Lloyd in the real picture and at the same time deliver a report of mine for transmission to General Cromwell.'

Donna suggested that Derek and John leave immediately for Castle Clarke. She warned Luke that Sir James would not have believed that Luke's mission was to escort Fenella to visit a sick relative. There was probably a large body of Highlanders about to descend on the hamlet. John, with his weapons restored, left with the prisoner. Donna brought all the horses into the barn as the weather worsened. They could not return to Greytower until conditions improved. Everyone sat around a large pot of beef and vegetables into which oats were progressively added. This hot potage was a welcome meal.

Luke and Andrew had just settled themselves down for the night in the barn annexe when the door opened allowing a blast of chilled air and fluttering snow to engulf them. Both men were on their feet with swords drawn as several Highlanders burst in waving their heavy swords in the air, and motioning the Englishmen to enter the main house. There Sir James confronted them. The Scotsman did not mince words, 'Explain yourself Colonel! What are you doing in this peasant hovel with known criminals and a hostage?'

'Exactly as I told Major Burns at Greytower. Lady Fenella received news that there was a sick relative here that the locals could not afford to maintain. It was not a relative, but Morag Ritchie whom Sir Derek and Captain Mackail had left here so that they could move more quickly away from you. We intended to return to Greytower this morning, but the conditions were too bleak. Morag's health had to be our first priority.'

Sir James was not impressed. One of his men entered the house and the two conversed in Gaelic for some time. Sir James turned to Luke. 'Where is Halliwell?'

Luke decided not to tell Sir James of Derek's surrender, as there would still be time for the Highlanders to overtake them. He told a half-truth, 'I ordered him to continue to Castle Clarke to deliver a report to the English high command.'

Sir James drew in his breath and then barked a few orders in Gaelic. He turned to Luke, 'We will see your party off in the morning. In the meantime my men and I will take up lodgings in the hamlet's other houses. When you have gone I will question the inhabitants, kill the stock, and raze the hamlet to the ground. These people have assisted convicted criminals. Good night to you all.'

As soon as Sir James departed an agitated Donna spoke, 'Colonel you and your sergeant must sleep in the house, and protect the women. I and my servants will sleep in the barn.' Deep in the night Luke heard his horse neighing and the village cattle lowing. Luke knew what was happening and pulled his blanket more tightly around him as some cold air from the barn invaded the inner chamber. By morning the snow had ceased and Luke's party left. As they made their way to the main road they saw Sir James riding in their direction. He raised his arm in salute. Luke smiled to himself. He could imagine the gruff Scot's fury when he discovered that the villagers with their livestock had fled during the night. As they climbed the mountain path Luke looked back towards the hamlet. Massive clouds of smoke billowed into the air and in the dull winter's light dozens of bright orange red fires were clearly visible. Sir James was a man of his word. Luke rode up beside Andrew who was leading the group. He spoke quietly to his sergeant who turned around and headed back down the mountain.

When Luke reached Greytower late on the following day Elspeth and Janet were delighted to see Morag, but concerned about her appearance. While the women fussed over the girl, Luke and David gathered in front of the blazing fire in the hall, and drank whisky as they brought each other up to date. David reported that he had kept the remaining guests under constant guard with two men allocated to each of Elspeth, Janet, Mungo and Malcolm.'

'And did you pick up any relevant information?' Luke asked.

'No, but close surveillance and continued captivity seems to be getting under Sir Malcolm's skin. He is agitated, demanding his immediate release and escort to Stirling. He was overtly concerned with Lady Fenella's absence,' replied David.

'What about Mungo?'

'Very relaxed, and he has suddenly appointed himself as Elspeth's protector.'

'In what sense?'

'Wherever she goes he follows at a discreet distance.'

'Odd, is he waiting an opportunity to harm her?'

'It's possible, but Lady Elspeth is never alone. Apart from my guards, Janet Hudson is her constant companion.'

Luke thought the time was right to clarify his position with David and to assert his priorities. 'David, we must reassess our mission. Return all the women to their families. And assuming that Mungo and Malcolm remain suspects I will transfer them to Edinburgh,' asserted Luke.

21

avid agreed, but only regarding the women. 'I want the men kept under Scottish control. They must not be sent to Edinburgh.'

'We can discuss that later, muttered a peeved Luke. He suddenly looked as if he had been struck by lightning. 'My God, what if active agent of the Black Thistle is a woman?'

'Then,' replied David laconically, 'we have double the suspects.'

'In addition Derek and Duff may still have a political agenda as well as their criminal activities,' sighed Luke largely to himself.

One of David's men entered the room. 'Major, a lone horsemen is on the trail heading down the mountain in this direction. Do we apprehend him or allow him to pass by?'

'Follow him until he is well passed the entrance. Do not alert him to our presence. The sleet should conceal us.'

Half an hour later the soldier returned accompanied by a heavily coated man covered in snow. As he pulled back his hood, Luke gasped, 'Cuds me, we thought you were dead. Welcome back Sir Alistair. Where in the world have you been?'

'Forgive me gentlemen; I must speak to my wife. I will explain everything in the morning.'

Sir Alistair left, and the two soldiers settled more deeply into the chairs that they pulled closer to the fire. Luke built it up with several pine logs and was soon asleep. He was awakened by the entry of one of David's men. 'Major, I must report a strange occurrence. Ever since Sir Alistair has returned Lady Elspeth has been crying, at times almost hysterically.'

Luke and David followed the guard back to the Stewart's chamber where the second guard quickly jumped to attention. There was no need to press one's ear against the door. The constant throb of someone crying could be heard well into the corridor. 'Her ladyship is certainly distraught, but we should not intervene in what is a matter between husband and wife. Sir Alistair has news that has upset Elspeth. We may be informed in the morning,' concluded Luke hopefully.

Early next morning before dawn, but after the castle was stirring with the bustle of servants stoking fires and preparing breakfast, Luke heard a gentle knock on his door. He struggled to his feet and opened it. Gillian Shaw was there. She spoke quietly, 'Lady Fenella wanted you to know that she could not sleep last night largely because you did not visit her, but also because she could hear Lady Elspeth crying throughout the night.' Luke stole a hug and kiss, and Gillian went about her duties.

At breakfast all were present except Elspeth. Alistair excused his wife's absence, and pre-empted any questions Luke and David were ready to ask. 'I brought Elspeth news concerning her family. She took it badly, and wept most of the night. She has just fallen asleep in the last half hour.'

Macdonald asked, 'And where Sir Alistair did you pick up this shattering news regarding her ladyship's family?'

Luke looked at David as both pondered on the temerity of Macdonald enquiring into the private life of a social superior. Alistair's eyes revealed his anger, and he ignored the question. Luke took advantage of Macdonald's breach of protocol. 'Sir Alistair, we are all anxious to know what happened to you from the moment you disappeared. Most of us assumed that you had been murdered, and thrown over the cliff.'

'I received a message to meet Sir Derek in a cave that was not far from the concealed door on the lowest level. I had trouble opening the door with its giant bar. It took me some time to work it loose.

And then I had to manipulate a stone head. I left the door open to facilitate my return. The blizzard was intense. Through the snow I saw a light flickering in the cave. I struggled to its entrance but it was deserted. The light came from a small candle which had obviously been lit by somebody but as it was about to go out I made my way back to the Greytower door. It was shut, and locked from the inside.'

The women around the table had stopped their trivial banter and were listening intently. Janet asked, 'Sir Alistair, how did you survive? You should have frozen to death.'

'I was lucky, Mistress Hudson. I returned to the cave and stayed there until morning when I noticed some light at its far end. It was a small exit which opened onto a narrow precipitous path which hugged the wall of the cliff and descended slowly into a tiny glen. I obtained food and additional clothing from the sole inhabitant, and eventually made my way over several days to our headquarters at Stirling.'

David eagerly asked, 'As you have been to Stirling do you have new orders for me?'

'Yes. Only a few of the great nobles are still resisting the move towards giving the King real power. Most scramble to become a courtier. Parliament is about to declare for the King and remove all restrictions on the Royalists. They now dominate army and parliament. Even the Kirk has removed its injunctions against those who had opposed the Covenant. I spoke to the Marquis of Argyle who straddles both camps. He ordered me to return here, and asked you Major to bring all the prisoners to Stirling.'

'What about the investigation into the Black Thistle and into the murders of Mackelvie and Caddell, and the attempt on your own life?' asked Luke.

'Those at the centre of power are no longer concerned with these minor issues. If the King is about to rule as well as reign then the Black Thistle's activities will be seen as meritorious. You have succeeded in your original mission to rescue the prisoners originally held in Castle Clarke.'

Luke was not impressed. 'This is not good enough. Our mission is incomplete, and I certainly have no orders to withdraw from this enterprise. I want the Black Thistle uncovered, and the murderers of Mackelvie and Caddell brought to justice. Have you written orders for Major Burns?'

'No, given the dangerous situation at the top even the Marquis was not willing to put anything in writing.'

David turned to Luke, 'You have changed your mind. I thought you decided that the murder of Mackelvie and Caddell was solely a Scottish matter. Now you want to use their demise as an excuse to stay involved when any reason for your continued participation in this mission is gone.'

'On another matter–Alistair, what news did you bring your wife that was so devastating?' asked Luke insensitively.

'That is a personal matter, and I am surprised that you ask it Colonel.'

'Let's not waste words on etiquette. You wife is a critical element in this whole situation. Something that disconcerts her in the way your information did, must have broader connotations, and relevance to our enterprise,' replied Luke.

Alistair thought for some time before answering, 'Elspeth was hoping to return to her father and have her baby in his home castle. The Earl of Barr is so distracted by events that he now wants Elspeth to go south to the Borders and stay with my sister while she delivers the child. Elspeth does not like my sister and refuses her father's request. She wants me to take her to Barr Castle immediately.'

Luke decided against further questions and Alistair returned to his wife. David was agitated. 'I don't like any of this. Why did Alistair receive verbal orders from Argyle when he knows this mission is under the control of the Earl of Barr? How does he have a personal message for Barr's daughter, yet nothing for me? In addition Elspeth is a strong-minded woman. The news he brought had to be much more shattering than the fairy tale we have just heard.'

'Given the anarchy at the centre of the Scottish government, and the pressure on you to bring the household to Stirling I can solve your problem. Hand everybody over to me. I will take them to Castle Clarke or Edinburgh. If they are all under English protection then whatever happens at Stirling will not affect them,' offered a scheming Luke.

The discussion was ended by a single shot that resounded through Greytower. The victim was quickly located. Lying across the top of a stairwell was a young Scottish officer. At the bottom of the stairwell was another body lying on its back, looking vacantly into space. Luke left David with the Scots officer who was stirring, and bounded down the stairs. The prone body dressed in black with an elaborate lace collar was the balding Mungo. Luke was impressed by the physique of the victim and was not surprised to find hidden under his clerical top coat several daggers and a pistol. Luke leaned over the body and ascertained that he was still breathing. Luke rolled him over and found a pistol wound through the shoulder blade. There was not much blood, and Luke assessed that the fall, and not the shot accounted for his comatose condition. Mungo was carried to his chamber, and Fenella tended his wound.

The Scots officer had fully recovered. He recalled that he was on duty outside Lady Elspeth's room when he walked to the top of the stairs to exercise his legs. The next thing he remembered was coming to, lying across the top of the stairs with a fellow officer shaking him. His pistol had been taken. The scenario appeared clear. Someone had hit the young officer, stolen his pistol and shot Mungo. David asked Dugall to remove the shot while the victim remained unconscious, but the scream that emanated from his chamber indicated that he had awakened before the minor surgery was completed. After some time he demanded to see David who asked, 'What happened?'

'Sir Alistair shot me.'

'You saw him?'

'No.'

'Then how do you know?'

'It's obvious. What has changed in the last few hours? Sir Alistair has returned.'

'But why would Sir Alistair wish to kill a minister of the Kirk?'

Mungo avoided an answer by falling asleep–or feigning such a condition. Luke, when told, shook his head, 'Alistair is innocent. He was with us until a minute or so before the shot. There was no time for him to prime your cornet's pistol. If Mungo had been stabbed Alistair might have had time. Secondly, Alistair is an excellent shot. He would not have missed vital parts from such close quarters. This is the work of an amateur not well versed in pistols.'

'Yet someone well enough versed to prime and fire the weapon?' David responded.

'Why would Mungo accuse Alistair?' asked Luke.

'He is obsessed with Elspeth, and wants the husband out of the way,' replied David.

'We suspect that Mungo is Barr's man, designated to protect Elspeth,' said Luke.

'Or murder Alistair? Maybe Alistair got in first?' David surmised.

'But he didn't have the time,' repeated Luke.

Luke and David found Alistair and the three men sat on the stairs of the shooting. Luke asked as innocently as he could, 'Alistair, what did you do when you left us?'

'I returned to my room.'

'Were the guards on duty?' asked Luke.

'The cornet was outside the door pacing up and down between it and the stairwell, and the trooper was inside the chamber.'

'Lady Elspeth was with you at the time of the shot?' continued Luke.

'I think so.'

'What do you mean you think so?' interrupted Andrew.

'Elspeth was here when I returned, but she passed me at the door, saying she was going to look in on young Morag. I don't remember where she was at the moment the shot was fired.'

'What happened when you heard the shot?' asked David.

'I was just inside my open door. The guard and I ran to the top of the stairs where we saw the body of the cornet.'

At that moment Elspeth was seen leaving Fenella's room. Luke called her over, 'Your ladyship, where were you when you heard the shot?'

'I had just entered Fenella's room.'

'What did you do when you heard the shot?'

'Fenella asked me to bar the door in case the assailant was after Morag or either of us. I did not open the door until one of your soldiers explained what had happened just a moment ago.'

David took up the questioning, 'Can either of you explain why a minister of the Kirk should accuse Alistair of shooting him?'

22

'**D**id he see me?' asked Alistair evasively.

'That would have been difficult since he was shot in the back,' interposed Luke.

'No,' answered David 'He did not see you, that is why I ask the question. Why would he assume that of all the people in the castle it was you who shot him?'

Luke noticed the colour drain from Elspeth's face when David mentioned Mungo's accusation. Before he could question her she fainted. Fenella and Janet were summoned. Most attributed the fainting spell to her pregnancy. Luke was not so sure. The Stewarts were up to something.

He interviewed the young cornet that had been knocked unconscious. The soldier described meticulously alternating between standing rigidly at attention, and patrolling the corridor fronting the bedchambers of the Stewarts and Fenella. The soldier explained, 'Just before I was hit I noticed a large rat at the top of the stairs. I crept towards in, and primed my pistol.'

'Good God man, why didn't you mention this earlier? Your pistol was primed, and you were ready to fire.'

'Yes.'

'Did you hear or sense anybody near you just before you were hit?'

'Yes. Mr Macdonald walked past me. I asked him to proceed carefully so as not to frighten the rat. He stopped part way down the stairs to watch me exterminate the creature.'

After dinner Luke and David re-examined the events of the day. David concluded, 'Given the cornet's evidence Alistair would have had time to leave us, strike the ensign on the head, grab his primed pistol and fire at Macdonald who was hovering on the stairs, and be back at his chamber's door to run out with the other guard.'

The men continued drinking well into the night and eventually David retired. As if she had been waiting for the Major's departure Gillian entered the room, and within minutes she and Luke were stretched out on the large fur rug in front of the fire. Their lovemaking was frenetic as if it was the last occasion for their mutual outburst of lust. The active thrusting and gyrating continued after short interludes of gentle caressing. Both parties eventually became exhausted and were reduced to whispering sweet nothings to each other. Luke was on the edge of sleep when Gillian said quietly, 'I know who shot Mr Macdonald.'

'Stop teasing Gillian! Snuggle up and sleep!'

'Don't you want to know who shot the minister?'

'Alright tell me what you know,' muttered a drowsy Luke.

'No, you are not taking me seriously. My evidence should not be dismissed as a joke, just because I am a servant.' With that Gillian rose and strode out of the room. Luke made no attempt to follow her. Just as she disappeared into the corridor she muttered, 'It was Lady Elspeth.' Luke jumped to his feet but by the time he had reached the door Gillian had disappeared.

It was mid morning before Luke found Gillian alone. Fenella had kept her busy as she had the added responsibility of looking after Morag, who was slowly regaining her strength. Luke made a half-hearted apology for not listening to her the previous evening, but excused himself on the grounds that their mutual activity had so exhausted him he was virtually asleep.

'So now you are ready to hear my evidence?' pouted Gillian.

'Yes.'

'I will only give my story to you if Major Burns is also present.'

Luke waited until David arrived and announced, 'Gillian accuses Lady Elspeth of shooting Macdonald'

David was shocked. 'You had better have proof, or I will arrest you for maligning one of your superiors. A servant accusing a gentlewoman of attempted murder is unacceptable.'

'Major Burns, you are as arrogant and demeaning of me as Colonel Tremayne. Do you want to hear my story or not?'

'Go ahead wench,' David replied without much enthusiasm.

'Just after the shot was fired I came down the stairs and collided with Lady Elspeth about to enter Lady Fenella's room.'

'So you didn't see the shot fired? Your evidence is simply that Lady Elspeth was in the same area as the shooting,' reiterated Luke.

'I am not daft Colonel. I have lived with soldiers long enough to know the smell of gunpowder. Lady Elspeth reeked of it.'

'If that is true Elspeth lied to us. She was not in Fenella's room when the shot was fired. It will be easy to verify,' concluded Luke.

Fenella, Morag and the guard on duty were questioned and all agreed that Elspeth arrived immediately after the shot was fired. Luke summoned Alistair and Elspeth. Alistair immediately protested that Elspeth's condition was delicate, and further questioning bordered on harassment. Luke insisted, 'Sorry Alistair, but new evidence suggests that Elspeth shot Mungo.'

'No, Colonel you have picked on the wrong Stewart. I shot Macdonald. He had been showing undue attention to my wife, and he was sent here to kill me. I did not intend to kill him. Just to warn him off. I am a very good shot. If I meant to kill him he would be dead, not nursing a minor injury.'

'Don't lie Alistair. The evidence points to Elspeth, and frees you. Why did you lie to us Elspeth?'

'When did I lie to you?'

'You said you were in Fenella's chamber when the shot was fired. Three persons affirm that you did not enter the room until after that shot was heard. Another witness could smell gunpowder all over you. You fired the shot, or was close to it when it was fired.'

'It's my fault gentlemen,' persisted Alistair. 'I led Elspeth to believe Mungo was an enemy, and in her condition, not thinking clearly she did what she did.'

Elspeth looked astonished, and turned on Alistair. 'What are you on about, my husband? Do you think I set out to kill Macdonald? Does my so called condition turn me into a murderous idiot?' Alistair tried to soothe Elspeth who pushed him away.

Luke stared at her and demanded, 'What happened?'

'I did lie about where I was when the shot was fired. I thought the truth might complicate the situation. Alistair did not shoot Macdonald, and neither did I.'

'So the only other person in the vicinity, my cornet, shot at Macdonald?' asked David.

'In a sense, yes.'

'Then who knocked him out?' continued David.

'Oh, I did that,' admitted Elspeth.

Luke raised his hand and quietly said,' My lady, tell us in detail what happened.'

'I decided to visit Morag. Noticing that we had an excess of candlesticks and remembering that Fenella complained that she had too few, I took one of ours with me. On leaving my chamber I noticed that the guard on duty was stalking Mr Macdonald. The cornet had his pistol primed and had it pointed at the stairs. Before he could fire at Mr Macdonald I struck him with the candlestick. He fell to the ground, but in the process his pistol went off and gunpowder spilt on my clothes.'

'Did the pistol fire when you hit the soldier or when it hit the ground?'

'I don't remember. Having saved Mr Macdonald, as I thought, I ran towards Lady Fenella's chamber and collided with one of her servants.'

'Didn't you see the rat? The cornet was about to shoot the rat, not Mungo. By knocking him out his pistol fired and the shot randomly hit Mungo who was on the stairs watching the rat being dispatched,' commented Luke.

David continued, 'Why did Mungo not tell us what had happened?'

'Ask him,' replied Elspeth.

They did, and he admitted, 'I saw Lady Elspeth hit the soldier and the next minute I felt a pain in my shoulder, lost my footing and fell down the stairs, knocking myself out. When I came to and thought about the situation I assumed what indeed has turned out to be the truth. Lady Elspeth in the darkened corridor had thought the soldier was about to shoot me.'

'Why did you not tell us the truth?'

'I have taken an oath to protect Lady Elspeth from harm. I am here to protect her, not to accuse her.'

'Yes, we note your desire to protect Lady Elspeth when her husband is away,' Luke commented sarcastically. 'If you are here to protect Lady Elspeth why did you accuse her husband of shooting at you?'

'I am bound by my oath to say no more, but if you were a Scot you would know most political, religious and social problems have their origins within a family.'

'And as a Scot, I have long learnt not to trust a Macdonald,' commented David.

'Is Lady Elspeth is in danger from her husband?' asked Luke.

'Ask why she has been distraught since his return,' Mungo replied.

'Good heavens, man. Her ladyship is pregnant and about to provide Sir Alistair with an heir that will transfer the wealth of the Barr family to him. He has very powerful monetary reasons to protect his wife,' reiterated David.

'Precisely my point, Major. Lady Elspeth is safe, only until she delivers an heir. After that not even your combined armies can save her.'

Luke and David returned to a small chamber that had become their headquarters when immediately Malcolm entered the room and protested, 'You were sent to rescue us. That was weeks ago yet we are still not able to leave.'

'You forget Sir Malcolm that my mission was not only to rescue you, but to uncover the identity of the Black Thistle. Since those initial orders we have had two murders that need to be solved, and I am uncertain as to the safety of the women under our protection. Someone tried to kill Morag, and then she was abducted. Many express concern for the safety of Lady Elspeth. Until I know who ordered their abduction to Castle Clarke in the first place I am uneasy about releasing them,' answered David.

'In addition,' added Luke, 'the state of Scottish affairs is such that if we did free you, where should it take place? I want to take you to Castle Clarke or Edinburgh currently held by the English, and free of the intrigue and conflict that surrounds the Scottish government at Stirling where Sir Alistair wants you relocated Stirling.'

Malcolm looked alarmed at both possibilities. He remained silent for several minutes. He then spoke quietly,' Gentlemen, I have a proposition to put to you. I know who was behind the kidnapping of Morag, Janet and Lady Elspeth. It was a personal matter—an aspect of a past grievance. It had nothing to do with the current political agenda. They could safely be returned to their families. Whatever plans the kidnapper had for those women were scuppered by your actions, and by the arrival here of persons that were not part of the kidnapper's plan. Sir Alistair and Mr Macdonald were not victims of the abductor, and may have been sent here to undermine the original kidnapper.'

'Just a minute Sir Malcolm. If you know why the women were kidnapped and that Sir Alistair and Mr Macdonald were not part of the plot then are you the kidnapper?' interrupted Luke.

'Not necessarily,' prevaricated Sir Malcolm.

'This is not a game. Who is the kidnapper?' asked an angry David.

'Given the turn of events on the national scene and the developments at Castle Clarke and Greytower it is not in anybody's interest to name names. All I will say is that the murders of Caddell and Mackelvie were also not part of the abductor's plot.'

Luke spoke quietly, 'The picture you paint Sir Malcolm is rather difficult to accept at face value. You suggest that the original kidnapping was solely a personal matter and with the expiration of time the kidnapper is now happy to let the women return to their families. You imply that the original plot was subverted by the arrival of Sir Alistair and Mr Macdonald and that the murders had nothing to do with the basic plot.'

'It's the truth. The other part of my proposition is that you allow me to return to my new estates on the outskirts of Edinburgh and to take Lady Fenella with me. During my stay here I have become very fond of her and with the desertion of her husband she cannot remain in the Highlands. She has indicated that she is willing to accept my hospitality.'

David and Luke looked at each other. The story that Fenella and Sir Malcolm had been seen together in an intimate embrace was obviously true. David answered, 'You have given us much to think about Sir Malcolm.'

Sir Malcolm left, and David placed another log on the dying fire and said. 'Luke, Malcolm has told us more than he intended. He admits that the original kidnapping of the women was part of a family feud. The kidnapping of Elspeth the only daughter of the second most powerful nobleman in Scotland, a man central to the Scottish Government, suggests that the feud is against the Earl of Barr. My special unit was authorised and funded by the Earl of Barr who never doubted that a political or personal enemy was behind the abduction of his daughter. It seems reasonable as you suggested returning Elspeth to her father and Morag and Janet to their families. They appear genuine victims. However Fenella, Alistair, Mungo and Malcolm remain suspects as one or more of the Black Thistle, the murderer and the abductor.' Luke agreed.

23

At dinner David publicly responded to Malcolm's proposals. 'First thing in the morning Lady Elspeth will be returned to her father, the Earl of Barr, and the other women to their family homes.'

Alistair jumped to his feet flushed with anger. 'No. My wife stays with me. I am her family, not the father who has disowned her.'

Elspeth nodded and said quietly, 'Alistair is right. You heard me crying throughout the night. My father no longer wants me at his castle.'

Mungo looked amazed, Malcolm bemused. David continued coldly, 'Your wishes are not relevant. My original mission was to free you Lady Elspeth, and bring you to your father. Those orders stand. Have you any evidence, other than what Sir Alistair told you, that your father has suddenly turned against you?'

'No,' replied a bewildered Elspeth.

Luke added to her unease, 'Lady Elspeth, does it not strike you as strange that within weeks of you delivering an heir to his vast fortune and power, that your father would reject you? The baby is central to the continuation of his power and wealth. Maybe Sir Alistair was mislead by relatives who would benefit from your failure to produce an heir, or the removal of your father from the child's upbringing.'

'Or more likely Sir Alistair's relatives who want to control your imminent child instead of your father,' added David with a touch of menace, and a penetrating stare at Alistair.

'My rights as a father are paramount, whatever the Earl of Barr, or my own relatives think,' shouted Alistair.

'Enough! The women will be ready to leave at first light,' reiterated David, determined to brook no opposition.

Alistair led away a weeping Elspeth, Fenella fussed over young Morag as she led her to her chamber, while Janet placed her head on the table and prayed before she eventually left the room. Malcolm alone remained, silently fuming at the soldiers. He ultimately exploded, 'You bigoted arrogant fools! Why not free Fenella? Why keep me here?'

Luke responded, 'Sir Malcolm, if this was Ireland I would imprison you, and then subject you to considerable pressure to reveal what you know. A diet of stale bread and water at your age, even without any physical abuse, would probably be sufficient incentive.'

'Major, are you allowing this English curmudgeon to threaten me?' Malcolm pleaded in the direction of David.

'Yes, until you reveal the name of the abductor, and all you know about the murders and your fellow guests here, a trip to Castle Clarke and subjection to English military justice may be exactly the medicine you need. The English have always used torture most efficiently.' Malcolm paled and stumbled from the room.

Next morning nine Scots troopers under Dugall's command, and three women left Greytower. The weather was bleak. The party would stay together until they reached the nearest eastern glen where they would separate. Elspeth would be taken to the south, Morag to the east and Janet to the north. Luke was surprised that Alistair did not farewell his wife. It was Gillian who provided an explanation later in the morning. 'After the women left Lady Fenella asked me to supervise the cleaning of all the bedchambers—so that some could be used by yourself and Major Burns. Last night Sir Alistair did not sleep in his bed.'

'Where were the guards?'

'Major Burns withdrew them to prepare for escort duty.'

'Maybe he was a little premature,' muttered Luke.

Luke casually informed David over the midday meal that Alistair had disappeared. The Scot exploded. 'Why didn't you tell me earlier?'

Luke taken aback by the outburst asked, 'Calm down! Why are you so upset?'

'I have met several stupid English officers but sometimes Luke you head the list. Alistair has disappeared to organize the rescue of his wife. At all costs he will prevent her returning to her father. We should have foreseen this given his passionate views on the rights of a father. I should have placed him under guard.'

A tense silence followed David's outburst. Luke paced up and down the chamber in a state of some embarrassment. Finally David quietly announced, 'I must warn the escort parties of Alistair's intentions, and send more troopers to reinforce those with Elspeth.' Luke, anxious to redeem himself, offered to lead David's troopers on the mission. After further contemplation David, implicitly accepting Luke's offer decided, 'A large force is not necessary. If the existing escorts are prepared for any attempted rescue they can cope. Three of my men will be sufficient for your party.'

Luke and three Scots troopers were quickly on the road. Luke wallowed in a modicum of guilt. He had delayed informing David because he had spent the morning with Gillian. Elspeth and her escort party were half a day ahead of them. Alistair probably had an additional twelve hours to plan and organize his rescue attempt. Luke's mood worsened. Hours of slow progress, now in darkness and heavy snow, added to his frustration. Eventually the mountain path descended rapidly and the snowfall ceased. In the crisp mountain air Luke saw the steady glow of a campfire in the glen below them. They had caught up with Dugall's party. When he reached the outskirts of the camp Luke fired his pistol, and his troopers shouted in unison to identify themselves. The sleeping soldiers were soon on their feet, and Dugall approached Luke, 'Greetings Colonel, what is so important that it could not wait until morning?'

'Major Burns sent us to reinforce the group escorting Lady Elspeth. Her husband plans to abduct her.'

Dugall shouted, 'Bring Lady Elspeth to the fire!' Luke had been surprised that the three women had not immediately joined them to discover why they were rudely awakened. In the firelight, well back from the flames, he saw Morag being comforted by Janet. Lady Elspeth was nowhere to be seen. He was uneasy. The trooper that went to fetch Elspeth came running back to the officers gathered around the fire. 'Sir, Lady Elspeth is missing. Her blanket has been placed over a couple of low rocks to give the impression that she was under it.'

Luke intervened, 'Trooper check the horses. See how many are missing.'
He soon returned. 'Sir, all the horses are gone.'

134

Dugall swore. 'I did not expect horse thieves in this location, or in this weather.'

'These were no thieves,' replied Luke. 'This was Alistair. He freed his wife, and then carefully led your horses away. Let's hope he does not take them far.'

'What do we do now?' asked Dugall.

'At first light you return with the ladies to Greytower. There is no point continuing on foot. Of the four horses that came with me, two can be used to carry the women and I and one of the troopers will use the other two to search for Alistair.'

'No, Colonel I cannot agree. I have orders from Major Burns to deliver these women to their families as soon as possible. I will accompany you on one of the beasts until I find our horses. I will bring them back here, and we will then continue to deliver the gentlewomen to their homes.'

This was no time to argue with Dugall who probably was correct in asserting that Luke had no authority to countermand David's orders. As soon as it was light Luke, Dugall and a trooper headed after Alistair and the horses. It was slow progress. Around midday when Luke heard it–the rumble of artillery fire. All three were amazed. 'Who has heavy cannon in this area at this time of the year?' asked Dugall.

'Not us,' replied Luke. 'We will not attack anywhere until late in summer.' He then had second thoughts as the noise reverberated around the mountains creating an awesome sound. ' On the other hand the Lord General is so besotted with his artillery that he may have brought some cannon into the area.'

Then all three froze. The sound got louder. It was not artillery. Luke and Dugall flung themselves against the cliff face that edged the path. The trooper was too slow. He was swept over the precipice. Giant ice boulders interspersed with loose snow and fragments of ice and rock bombarded the path. The avalanche lasted several minutes, taking away most of the path behind the soldiers. A man might make his way carefully back up the mountain, but a horse could not fit along the few inches of track remaining. Both survivors were shaking. Luke was so shocked that he was unable to speak.

A pessimistic Dugall was the first to comment. 'Even if we recover the horses in the next few hours there is no way we can get them back to the camp along this track. It will take days if not weeks to find an alternative pathway. I will return to camp on foot and explain the situation. The women will have to go back to Greytower. Luke will you continue the search for the horses?'

'And Sir Alistair,' answered Luke.

Luke tied a rope around Dugall and entrenched himself behind a massive boulder around which he had anchored the rope. Luke perspired freely as Dugall eased himself slowly along what was now a ledge, only inches wide. The silence was chilling. It was suddenly broken by the sound of horses coming down the path. Luke desperately tried to prime his pistol. Dugall simply stopped where he was, an easy target for any enemy. Luke's concern evaporated as he recognised that the newcomers were speaking English.

Two horsemen emerged through the mist just avoiding the gap where once there had been a path. It was Andrew and John. A consensus was reached. Dugall would continue across the gap, his task made easier by having a person on each side to help him. With John he would return to camp riding Andrew's horse. Andrew would ease his way across the gap and on Dugall's horse accompany Luke in pursuit of Alistair and the horses.

Some time later as they cantered confidently down the lessening incline Luke asked Andrew, 'Did you have any trouble carrying out my orders?'

'No, I reached Castle Clarke before John began his return journey. It was crammed with English soldiers, unloaded from several ships. The loch was crowded with at least three of our warships and two transports. The new soldiers are to stay there until the spring offensive. Harry has things running very smoothly. He readily put together a troop of an additional twenty men for you, and is aware that you may need more. He already knew that Sir James had changed the balance of military power in the area and his regiment vastly outnumbered those of our joint Scots- English mission. Harry also recognized that while Sir James currently represented the Scot's government of the Earl of Barr he was by conviction an out-and-out Royalist and would soon be acting for the King, if he wasn't already.'

'Where are my men?' asked Luke.

'Half remain at Greytower with Major Burns, and the rest came to the camp with us.' During the afternoon the weather cleared further and Luke could see they were on the edge of a long glen dominated by wide loch. He could also see wisps of smoke that indicated that the glen was occupied. Luke's immediate problem was to determine which side of the loch Alistair and Elspeth had taken.

Then he saw it, a dainty handkerchief made of the finest material with lace trimmings. It was caught in a bush at the commencement of the left hand track, Luke was ecstatic, 'Elspeth is leaving us clues. She wants to be found.'

They quickened their pace around the loch but after two hours both were frustrated and cross. This side of the lake was completely uninhabited, and the track, hardly visible to begin with, now petered out into sheep trails that led back into the hills. Andrew was his usual unsympathetic self. 'Alistair wins. How could we fall for such a simple trick? He leaves his wife's handkerchief at the beginning of the path he is not going to follow.'

In the clear Highland air the glimmer of a single candle was visible for miles and Luke assessed that there were several inhabited houses on the other side of the loch. The arrival of two English soldiers after dark might not be well received by this isolated community. Instead of seeking their hospitality Luke and Andrew spent the night wrapped in their blankets pressed tightly against a narrow wall that protected them from the crisp breeze, and the gentle flurry of snow.

In the morning the two men approached the largest of the huts. Luke hammered on the door. A giant of a man, with long brown hair almost to his shoulders, and a large curved sword hanging from his waist, opened it. Any attempt at conversation was useless as the man spoke only Gaelic. But he soon made it clear by an astute use of his hands that the two Englishmen were invited to breakfast. The meal was the usual range of oat-based food with the addition of dried and salted fish which had undoubtedly come out of the local loch. There was no antagonism, and Highland hospitality surpassed itself with a draft of whisky to finish the meal. Luke and Andrew were relaxed. They did not notice the several men were gathering outside the house.

Luke made an attempt through signs to ask the householder whether a wealthy looking lady and gentleman had passed through the hamlet. Luke was surprised at the positive result. The Highlander pointed to Luke, pointed to his wife and held up two fingers and by running his fingers across the crude bench he was suggesting that they brought with them a large number of animals. He then ran his finger across his mouth. The gentleman and his lady had given the hamlet horses in return for their silence.

As Luke and Andrew left the house several armed men blocked their path, ensuring that they could only progress back the way they had come. They made no attempt to attack the Englishmen but grabbing the horses' reins the animals were pointed in the direction of the mountains. They would not be allowed to follow Alistair. As they trotted away from the crowd a general shout went up and the blockading group broke up in consternation. Everybody ran towards a lonely figure staggering up the path towards them.

Luke and Andrew turned about and enthusiastically joined the crowd as their former host welcomed a limping, blood covered figure. It was Alistair.

24

Alistair had his wounds dressed while he recounted his adventures. It took ages as he explained his activities in both Gaelic and Scots.

'When I returned from Stirling a few days ago I brought my wife some very devastating news. I lied to you earlier, Luke. It was not about the baby. It was that her father had ordered my assassination.'

'Why would your father-in-law want you murdered?' asked Luke, pretending ignorance.

'I have served my purpose. I have sired him an heir whom he will bring up himself. He wants Elspeth tightly under his control until she gives birth, and then the child will be taken from her and brought up by the Earl of Barr himself. That is why I could not let you take her to her father.'

'So Elspeth did not wish to go to her father, and willingly absconded from the camp with you the night before last?'

'Yes?'

'So where is she now?' asked Andrew.

'I wish I knew.'

'Alistair, how do you lose a wife?' asked Luke chuckling at the humour of his question 'Last night we argued. Elspeth changed her mind and insisted that I take her to Stirling to confront her father. I refused, arguing that she would be detained, and I would be killed. I demanded as her husband that she come with me to the protection of my family and kinsmen. She pretended to agree, and we headed off at dawn. Later in the morning we reached the far end of the glen and stopped to water our horses. Without warning Elspeth, who given her condition usually needed assistance, somehow remounted her horse and headed off. I threw myself in front of her horse. It reared, knocked me to the ground, and stomped on me as it took off. It rendered me unconscious. When I came to, Elspeth had long gone, and my own horse had wandered off. I knew there were many horses in this hamlet–fine cavalry steeds which I had freed from your camp.'

Alistair, his wounds dressed, ate a hearty meal and was in no hurry to follow his wife. Luke reminded him, 'Alistair you don't want Elspeth to get too much of a lead.'

'Not a problem, Luke. When her horse trod on me it damaged itself, and given her condition, horse and rider will not get far.'

'Do you know which way she is going?'

Alistair looked uneasy and confessed, 'No. She could take one of two paths out of the glen depending on whether she is heading east to her father at Stirling or south to other kin.'

'Which would she take?'

'I tried to convince her that her father wanted to kill me and take control of his grandchild, and that in this campaign he was supported by many of her relatives who did not want the transfer of Barr land and power to the Stewarts. She continues to trust her father. Nevertheless I may have sown enough doubt for her to seek sanctuary with other friends or relatives who have no political agenda.'

'Andrew and I will come with you,' announced Luke.

Alistair was not happy, but had no option but to agree. When they reached the end of the loch the path divided. Alistair and Andrew took the eastern track leading to Stirling while Luke took the southern path. They agreed to follow their respective trails until sunset. If they had not caught up with Elspeth by then they would return to where they were.

Luke's path skirted the mountain. He moved through a low pass and was soon in undulating country following a rapid and widening stream that had carved a path through the extended valley. The heavy snow of the upland regions gave way to a light dusting, leaving a thin covering along the path with occasional deeper patches against the walls. The valley contained large fallow fields with substantial houses adjoining the road that itself bordered the river. At the entrance to the first house he saw a horse, an English cavalry steed, tied to a post. Elspeth must have needed assistance earlier than she expected. He trotted up the curving drive to the main door. Before he could knock, a dozen servants surrounded him. He was apprehensive. A tall well-armed lad approached him with outstretched hand and a friendly smile. 'Welcome stranger, why are you here?'

'Lad, I am looking for a gentlewoman who is fleeing those that might wish to harm her, and her unborn child. I saw her horse at your entrance. Could you ask her and your master to receive Colonel Tremayne?'

Eventually Luke was led into a chamber off the great hall where he was delighted to see Elspeth sitting on a bench beside a well- dressed, mature Scots gentlewoman. Standing behind them was an older, grey haired male dressed in a deep blue doublet and matching breeches. A small lace collar and equally tiny cuffs accompanied his white shirt. He signalled for Luke to sit and introduced himself, 'I am Sir Alan Campbell. At this time I do not willingly give hospitality to an Englishman, especially a soldier in those regiments of God despising sectarians...'

'Enough Alan! The Colonel does not wish to hear your religious views. I am Jean, Lady Campbell. I welcome you as a friend of Lady Elspeth who has explained her situation. It is a terrible circumstance when you do not know whom to believe, your husband or your father, especially when you are so close to childbirth.'

'Elspeth, I am surprised to find you here. Your husband was certain that you were heading to Stirling to confront your father. He and my sergeant are looking for you along the eastern road,' Luke explained.

'That was my intention, but I was increasingly overcome with doubt. What if Alistair was correct and my father was out to control my child, and dispense with its parents? I decided to head south and place myself under the protection of the Marquis of Argyle. My pains suddenly increased and I sought hospitality at the first house I came to. By sheer chance they are Campbell kinsmen of the Marquis.'

'Fate was certainly on your side.' Luke replied. He then noticed Elspeth's red face and heavy breathing and asked, 'What exactly is your condition, my lady?'

'I have felt severe pains ever since I left my husband. I feared I would deliver alone on some mountain pass. I was so relieved to reach this house and to find Sir Alan and Lady Jean to whom I explained my situation.'

'After your child is born and you have sufficiently recovered I will take you both to a safe place,' Luke promised.

Sir Alan interjected, 'No, she and the child must stay here indefinitely. Neither the supporters of Barr or Stewart would risk upsetting the kin of the Marquis of Argyle.'

'I am sorry to disagree with you Sir Alan. Too much is at stake. Both sides would risk the consequences. Mother and child could be abducted from here with ease, and you would not know whom to blame.' Elspeth brought the conversation to an end. She indicated that her time had come. The men left the chamber and Lady Campbell called in a multitude of maidservants and sent for the local midwife. Luke was disconcerted by the screams he heard coming from the birthing chamber, and after an hour and a half he heard the crying of a newly born baby. Lady Campbell emerged and announced that mother and son were well, although Elspeth had lost a lot of blood, and did not appear capable of feeding her baby.

'How will it survive?' asked Luke naively.

Jean smiled, 'I do not know the custom in England but here wellborn women do not feed their children. One of our servants who just had a miscarriage will wet nurse the child.'

A week later Elspeth, Luke and the Campbells sat down over dinner to review the situation. Luke suggested that he escort Elspeth and the baby with its wet nurse to safety in Edinburgh Castle.

'Scotland is full of feuding families, even more divided now as how to react to the King, the Kirk extremists and the English. In Edinburgh Castle, under the personal protection of General Cromwell, Elspeth and son will be safe until she decides what she wants to do, and the truth of the conflict between her husband and her father is resolved.'

Elspeth was forthright, 'No, I am returning to Greytower, and to my husband. I need to find out for myself what is happening. I will not leave my baby James with anybody.'

Luke was also firm. 'My lady if you return to Greytower you cannot take the baby.'

Sir Alan was agitated, 'I agree with the Colonel. You cannot risk the bairn. Partisans of either side could overrun Greytower and abduct baby James. And now that you have delivered an heir your own life counts for nothing. You both must go to one of Argyle's fortified castles. My wife will accompany you to put our family in the picture.'

'Forgive me Sir Alan but the Campbell clan has many persons with views that do not coincide with that of the Marquis and yourself. I am sure that if Barr or the Stewarts knew where Elspeth and the baby were, they would bribe the servants if not members of the relevant household to facilitate an abduction,' conjectured Luke.

Jean quietly said, 'There is a compromise. Colonel, you escort the baby, its wet nurse and myself to Edinburgh, Alan will take Elspeth back to Greytower.'

Elspeth was not appeased. 'No, no, no. The baby stays with me. I do not agree with any of you.' As she shouted her defiance Elspeth broke into loud hysterical sobbing and involuntary shaking.

Luke put his arms around her and held her tightly. Her condition gradually mellowed into a gentle sobbing. Luke led her to a bench and sat beside her and quietly underlined the dangers confronting her and her child. He emphasised that keeping mother and son together would make the task of the evil minded so much easier. Eventually he felt he could broach the key question again, 'If you did accept Lady Jean's proposal what would you tell Alistair about his son?'

'I will not accept the plan, but if I did, I will say that I had a miscarriage and have spent a week or so recovering.'

'No, Elspeth too dangerous,' cautioned Jean. 'You do not know how your husband will react. Stuff some clothes into your skirt and pretend you are still pregnant. That will protect you for a few more weeks. At the right moment you can tell him the truth. After all he is the father, and has powerful legal rights.'

'That is why the baby should be out of his reach until the situation clarifies,' added Luke.

Elspeth started to cry again, 'I cannot leave my baby, or my husband. I hear what you say, but we will not be separated.'

Sir Alan and Luke left the room and after a brief discussion agreed on what had to be done. Early next morning without Elspeth's knowledge Luke, Lady Campbell, the wet nurse and baby James left for Edinburgh. Mid morning a distraught Elspeth escorted by Alan and half a dozen of his servants headed back into the mountains.

It was a bleak early spring morning that found Luke pacing one of the ante chambers of Moray House, General Cromwell's Edinburgh residence.

This was situated halfway between Edinburgh Castle, and Holyrood Palace in whose grounds most of the English troops were encamped. Luke paced, not because he was anxious, but because of the cold. Eventually an orderly signalled for Luke to enter the General's chamber. Cromwell was sitting behind a large desk coughing and clearing his large red nose with a voluminous handkerchief. 'Sorry Luke, I am not well. I have been in bed for weeks and our offensive against the Scots army will be delayed. What's the story behind the baby?'

Luke explained, and Cromwell concluded, 'Congratulations, you have accomplished your mission. I now need you here as the situation within the Scot's government worsens, and the King asserts his authority across the countryside.'

'No General, my mission is not complete. I have rescued three women who were kidnapped, although one has returned voluntarily to Greytower. I have two original inhabitants of Castle Clarke whose role I have yet to determine – a devious cleric who calls himself Macdonald, and Sir Malcolm Petrie a long time lawyer and diplomat, and servant of the Scottish Government. Two other inhabitants were murdered and their deaths have not been solved. There is also the role of Sir Alistair Stewart whose political and personal agenda calls for investigation. In addition you are aware that there is a large Highland regiment of Camerons moving across the Western Highlands who will declare for the King at any moment?'

'I trust I do not detect a tone of impertinence in your responses, Colonel. I am well aware, largely through your reports, of the situation. Consequently I am in the process of reinforcing Castle Clarke with several regiments of troops who will be ready to be deployed when my offensive begins, a process in which I want you to play a major part. We will be more than a match for the Camerons.'

Luke appeared not to hear, or perhaps rather did not comprehend, the General's reference to his future. He was obsessed with unfinished business. 'There is also the fate of another devious villain whose political stance is unknown. Duff Mackail is a brigand whose has the charisma to raise local criminals and cause considerable trouble. And then there is the identity of the leader, or still active member or members of the Black Thistle which I have yet to uncover.'

'The Black Thistle is redundant. Whatever power it wanted to assert through abducting several people no longer applies. As for Mackail, and the murderers of the two victims leave them to the Scottish authorities. Let Cameron deal with the case. He has the authority and the will. I repeat, your mission is completed.'

25

'**B**efore I outline your new tasks let me comment further on the mission you have just completed.'

'How anything meaningful was accomplished from that remote tower house surprises me,' commented Luke delaying a direct confrontation with his commander.

'Greytower was not so isolated as you think. You sent me several reports and our agents at Stirling have picked up that both the King and Barr have been kept up to date with developments at your isolated haven. There is no doubt that both parties had agents there.'

Luke sensed he might have found a short cut to one of his immediate and personal problems, 'Did your agents discover if Barr ordered the murder of Sir Alistair Stewart?'

'It is well known at Stirling that Barr and the Stewarts hate each other, and the King is happy to exacerbate matters. Their dispute is longstanding primarily over property and power, not current political allegiances.'

'Surely current alliances cannot be ignored. Barr has been a leading pro-Kirk anti-King nobleman, who in earlier times was enthusiastic about the alliance with the English Parliament against Charles I. The fortunes of his faction have faltered. Alistair's family have been Royalists since the late sixteen thirties when unlike the Presbyterian families, the Stewarts were willing to accept the Episcopal system of Charles I,' responded Luke.

'Enough of Stewart! The most important piece of news that you sent me was that one of the prisoners was Sir Malcolm Petrie,' said Cromwell.

'How could that boring old fogy be of interest?'

'Just before he disappeared Petrie had been on a mission for the Scottish Government. He was in Amsterdam attempting to raise money and men to assist the Scots. We fear he was successful, but he never reported back to his government. He was apparently kidnapped before he could do so.'

'Who would benefit from the failure of the mission other than ourselves?'

'The King. If the Scots army received funds and men from the Dutch Calvinists it would reduce the need of the Kirk party to rely on the King. Without Dutch help the Scots government is forced into the arms of the King. With Dutch help Barr could have kept Charles as a figurehead.'

'What if Petrie was raising Dutch funds for the King, and not the Argyle-Barr government?' Luke asked.

'Petrie is such a trimmer that there are those that believe such a scenario. But whatever his past Petrie can be won over to our cause. Any Scot who has extensive property in the Lowlands, especially in the Borders, must see the advantages of supporting us against the King. You are free to offer him concessions to facilitate such an outcome. Now regarding the Stewart baby. I have informed the Marquis of Argyle that we have the baby under safe custody under the care of his cousin. Lady Campbell, the wet nurse and the baby are ensconced deep within the Castle, and your old deputy from Dublin Castle, Captain Cobb, is to provide the tightest security.'

'Cobb, what is he doing here? The last I heard of him you sent him to the Americas with information for me on the Chesapeake. I thought he had returned to Ireland.'

'He had. However as soon as I captured Edinburgh Castle I needed officers experienced in the administration of large castles. Cobb was my man. He also believed that amongst a large number of defeated Irish royalists who were being recruited in Spain for service in Scotland was his nefarious brother. A defector revealed that they would soon be embarking for western Scotland to assist the Royalist cause. Cobb is determined to bring his sibling to justice.'

Luke could not contain himself any longer, 'General, let me return to the Highlands and tie up the loose ends.'

Cromwell rose to his feet and paced up and down the chamber. He rarely faced defiance from his senior officers. At last he spoke, 'You have been loyal to our cause for a long time. You must feel deeply to challenge me. You can return to the Highlands eventually, but not to tie up loose ends.

In fact I have two related tasks for you. The first but most difficult is to put pressure on the Earl of Barr. If he is to survive he needs to throw his weight behind the King, or seek an alliance with us. You are to use your friendship with his daughter to persuade him to join us. Failing that, you will discredit him in the eyes of the King. Timing is critical. There is no point in winning over a has-been. His time may have already passed.'

'Win him over, or destroy him,' Luke repeated rhetorically.

'Yes, more important is your second and immediate task. The Western Highlands are critical as the Scottish government increasingly involves the leadership of the King. I am concerned that before I begin my offensive against the Scottish army, fanatical Royalists gathering in the Highlands will heavily reinforce it. And they will be strengthened by thousands of Irish mercenaries.'

'Are your agents sure of this rumoured Irish assistance?'

'Yes, their departure from Spain is imminent. In fact the Spanish government, which is seeking support from the English Republic, has ordered them out. They will sail to the west coast of Scotland maybe by keeping well out into the Atlantic until they can move east across the north of Ireland. That is why I reinforced our garrison at Castle Clarke and sent three men-of-war to the area to destroy any reinforcements before they can reach their destination. The local contact for this Royalist mobilisation is, as you know Sir James Cameron. His clan lands contain many inlets in which ships could unload troops without harassment. There is a severe shortage of English cavalry in the area. Over the next few weeks Luke your task is to raise several companies of cavalry. I will give you my final orders when the new companies are ready.'

The General terminated the discussion. Luke rode up the hill to the castle. The corridors were patrolled, and each door guarded by troops in the familiar colours of a Dublin infantry regiment. Captain Cobb was doing his duty. Luke was challenged at several points by the guards, but the orderly who escorted him pronounced the requisite password. After much delay Luke found where the baby and its entourage were located. The wet nurse answered his knock. She led him into an inner room where she announced that Colonel Tremayne was here to visit Lady Campbell. Jean Campbell emerged from behind a four-poster bed and her face lit up. Luke noted that the crib was strategically placed beside the bed. After several minutes of small talk Luke was about to leave when Lady Campbell began to sob, 'I am sorry Colonel. I don't think I can do this. I hate being locked away in this castle. That blond headed Captain Cobb says I cannot visit friends around Edinburgh as I may be the target of abduction as much as the baby.'

Luke replied, 'What can I do?'

'Escort me around Edinburgh for an hour or so. I have not been here for years. You must have the authority to overrule Captain Cobb's restrictions.' Before Luke could reply the wet nurse entered and announced that the very same Captain Cobb was at the door. The two soldiers embraced each other enthusiastically. Cobb invited Luke join him in his chambers for a drink, and to catch up on old times. Luke apologised to Lady Campbell, but promised to return, and provide her with an escort.

The men settled into a drinking session in which they adversely compared the local strong spirits to those they had become used to in Ireland. Each brought the other up to date on their exploits since the heady days of the Irish conquest. Luke intimated his intentions regarding Lady Campbell. Cobb initially resisted, but conceded that as the baby and its entourage had only arrived a few days earlier, no one knew that they were in the city.

Next afternoon the temperature was unseasonably warm. Lady Campbell as befitted a gentlewoman of strong Calvinist beliefs was dressed almost completely in black except for the white lace collar that was evident beneath her heavy black overcoat. They left the castle and walked down the main road as gusts of wind swirled about them. Luke began to feel uneasy. His anxiety grew and he whispered to his companion, 'We are being followed. At the next gate turn in and wait for me several paces along the path.' Lady Campbell followed instructions while Luke hid beside the gate. In less than a minute two armed soldiers came through the portal. Luke jumped out and confronted them with sword drawn and his left hand on his dagger. The enemy had good intelligence. It was an anti-climax. Luke immediately recognised the men as part of Captain Cobb's unit, and they quickly explained that Cobb had ordered them to follow at a discreet distance. Luke found Lady Campbell standing behind a tree further up the drive, and the party slowly made its way back to the castle.

Luke spent weeks recruiting and provisioning several companies of cavalry. Provisioning for such a force required a combination of diplomacy, negotiation and a liberal use of Cromwell's authority to obtain the requisite amount of biscuits, bread, salted meat, oats, beer and whisky from the grasping merchants. Cromwell's final orders were a pleasant surprise, and promotion. Not only was Luke to lead the newly raised cavalry into the Highlands to counter Sir James, and any Irish reinforcements. He was to assume command of all English land forces in the area and co-ordinate the movement of English warships off the west coast of Scotland. His headquarters was to be at Castle Clarke. Luke relished this return to pure military activities.

Luke and his large party reached Greytower through a different combination of glens and mountain passes to avoid the mountain trail destroyed by the avalanche. English soldiers manned the outer defences of Greytower and Andrew, not David, greeted him. Luke quickly ascertained that both Alistair and Elspeth had returned, and had spent most of time together, appearing only for meals. Within half an hour an obviously angry David appeared. 'I am glad you are back. Your sergeant unilaterally took over Greytower, and consigned my men to menial tasks.'

'I am sorry David, but he received orders to consolidate English control in this area to forestall an expected Royalist attack from Sir James Cameron and invading Irish mercenaries.'

'Yes, but my unit is not Royalist.'

'No, but you have not been honest with me David.'

'What do you mean?'

'While you technically represent the Scottish state in the current situation that is a moveable feast. I know that when the crunch comes you do not represent the Scottish state, but the Earl of Barr. I understand you are married to his niece and part of his extended political family. I cannot be sure how you would respond to the issues still confronting us, especially regarding Alistair and Elspeth. Alistair believes the Earl of Barr has ordered his death. How do I know you will not be the executing agent?'

'I have no such orders,' protested David.

'I believe you, but the Earl may have got to one of your men, or has sent someone unknown to you to do the deed– Mungo or Malcolm for example. Did you not seek new orders from Barr some time ago? I would do nothing if I were you until your authority is renewed, and your mission clearly defined. Until then I cannot trust you to be pursuing the same objectives as I am.'

'Then Luke our joint mission is at an end. My men and I will leave for Stirling in the morning,' declared David. With five times as many English troops at Greytower than Scots, and hundreds more on call from Castle Clarke he had no other option. Luke decided not to raise the decapitation of Angus Mackelvie. He would eventually discover whom Barr had designated as the real contact.

Greytower was too small and isolated for his newly augmented regiment of horse and Luke set up a temporary base in a nearby glen where a row of deserted houses provided the nucleus of an effective camp. Snowfalls were declining and fresh young grass was appearing beside the river that flowed through the valley. Here was plenty of space and food for the horses. Luke denuded Greytower of all unnecessary troops and created a simple network of warnings should it be subjected to attack. A large pile of dry timber placed on the Greytower battlements would be lit and the smoke which would be visible from the new cavalry base would quickly elicit a response.

Luke called on Alistair who thanked him for finding Elspeth, and not taking her to her father. Luke responded, 'That was her own choice. She asked me to make enquiries into the accuracy of your claims, but was determined to rejoin you until the matter was clarified.'

'And have you clarified the situation? I gather you went on to Edinburgh.'

'The Barr family is out to destroy you.'

'I told you so.'

'But whether the Earl himself is involved is unclear. In any case Edinburgh gossip regarding the politics at Stirling cannot be relied on. How is Elspeth? Her delivery must be soon?' commented a disingenuous Luke.

'Strange she doesn't seem any bigger now than a fortnight ago. She claims that her dates must have been wrong. She is very fragile, bursting into tears at the slightest criticism. She has nightmares in which soft crying is interspersed with terrifying screaming. She talks in her sleep. She mentions a name Jean often, and I have heard her call out "Luke."'

'Lady Jean Campbell was the woman who took her in and nursed her back to health after the exhaustion of her sally into the wilderness. I found her at the Campbell's and was the only familiar face during this experience. She is reliving the trauma of her travelling. Now that Greytower is under English military control Elspeth and yourself should consider yourself prisoners. When Elspeth delivers, and decides what she wants to do I will reconsider your position.'

26

'**U**ntil our child is born, we accept the situation. And I am glad that Burns is leaving. He is married to the Earl of Barr's niece and I am not certain as to his loyalties. Nevertheless Barr will still have an agent here–probably Mungo.'

'Alistair, I do not fully accept your claim that Barr has ordered your execution. You could have created this element of fear so that your wife will stay with you, and as a consequence you can control the upbringing of your child. Surely, given the time you have been at Castle Clarke and Greytower, an assassin could have finished you off ?'

'You forget Colonel that an attempt has been made on my life.'

'Not that I recall. You were locked out of the tower house with a convenient warm cave and escape route, which enabled you to report to your master at Stirling. How were your interviews with the young lad who claims to be King?'

Alistair responded proudly, 'I do not conceal that I take my orders from the King.'

'And that is why I am confining you to Greytower. You may have been sent back here by the King to co-ordinate a Royalist force that will combine Sir James Cameron's troops with shiploads of Irish mercenaries arriving from Spain. That is why we are filling the glens with English troops, and the sea lochs with English warships.'

The door burst open. An English trooper, wearing a helmet which obscured much of his face, and holding a primed pistol in each hand, strode into the room. He fired both weapons directly at Sir Alistair who slumped to the floor. As Luke moved to assist the wounded man the assassin retreated unimpeded. Within minutes the room was filled with English soldiers. Luke asked whether they had caught the assailant as he stuffed torn fragments of his fine shirt into Alistair's wounds. The cornet that led the group looked confused.

'Your orderly said the shooter escaped through the window. I have sent men outside to track him down.'

'I don't have an orderly. Did you notice anything peculiar about him?'

'Well I was surprised he was wearing his helmet.'

'Did you recognise him?'

'Not with the helmet covering his face.'

Luke sent for Fenella's cunning woman who immediately staunched the bleeding with handfuls of sweet smelling soft herbs she had in her large basket. Yarrow replaced the blood soaked fragments of Luke's shirt. Luke put his face against the victim and indicated that Alistair was breathing strongly. Luke informed Elspeth who asked that Alistair be carried to their chamber. John Halliwell, who had treated a few battlefield wounds, assessed that one shot had entered the chest and had to be removed immediately or infection would set in. The second had grazed his head and ricocheted away, and apart from bleeding profusely, that wound was not dangerous.

Alistair was still unconscious so John, with litres of whisky ready to administer to wound and patient, used a thin dagger blade to isolate the shot. He removed it with a pair of tweezers. Blood started to flow once again and the soft herbs of the cunning woman failed to staunch the flow. Luke and John tore up another shirt after pouring some of their finest whisky over the wounds. With Alistair lying peacefully on his back Elspeth finally broke down and sobbed uncontrollably. Luke put his arms around her, and she shook, overcome with guilt, 'I never believed Alistair that my father wanted him killed.'

'My lady, there is no evidence that your father is behind this, but someone certainly resents the movement of the Earl's property into Stewart hands.'

'Will Alistair survive?' Elspeth meekly asked.

'Hard to tell. Infection is the greatest killer with pistol and musket wounds. The shot did not penetrate any vital organs; otherwise John could not have removed the shot without killing him. It will take days until we know for sure. In case the attacker tries again I am posting two men in the room and two outside the door.'

Luke gave Elspeth a big hug, and ordered Andrew to coordinate the attempt to find the would-be assassin. Half an hour later he reported back, 'We have not found the gunman, but we did find one of our troopers on the lower level. He had been knocked unconscious, left in his underclothes, and his weapons taken. The potential assassin is not one of us. He simply disguised himself as such.'

'But there is no one here except us, and Mungo and Malcolm.'

'Exactly.'

'Search their rooms, and bring them here!' ordered Luke. Mungo claimed he was in his chamber, and that the guard on duty could vouch for this. Unfortunately the guard had left his post to relieve himself at the critical time, and later at the change of guards the two soldiers chatted for some time with their backs towards the door. Mungo could have slipped out and returned without being seen. The search of his room revealed nothing. Luke suddenly grabbed Mungo's hands and sniffed. 'I detect a whiff of gunpowder.'

'Yes, you do,' Mungo readily admitted. 'When I heard that Sir Alistair had been shot I recovered my own pistol from the chest in my chamber, and checked that I had shot and powder. Some of it rubbed off on my hands.'

'That's true Colonel,' interrupted one of the soldiers. 'There was a pistol, shot and powder on a small table in his room. The pistol has not been fired recently.'

'Clever move, Macdonald. Kill Alistair with an English trooper's pistol, and then play with your own to cover your activity,' observed Andrew.

'Why would I, a minister of the Kirk kill a kinsman of the King?' responded Mungo.

'Because you are not Mungo Macdonald, and in recent times not a clergyman. You are an agent of the Earl of Barr sent here to kill Sir Alistair,' Luke pontificated.

'You are absolutely correct in one regard. Yes, the Earl sent me here when he discovered how to infiltrate the system devised by the Black Thistle, but my mission is not to kill Sir Alistair. It is to protect Lady Elspeth. The Earl feared that his heir and her child would be the target of her husband's relatives.'

'If the Earl sent you to protect his daughter, why did he also send her husband?'

'He didn't. The Earl was content to leave her here with my protection, especially after Major Burns arrived. I knew he was a special agent and close relative of the Earl. Neither Barr nor Burns sent Sir Alistair here. That gentleman came here of his own accord, or on orders from the King.'

Malcolm was cross, as usual. He had been asleep and did not hear the shots, or the ensuing furore. When the troopers were finally admitted to his room he thought he was about to die. When the matter was explained Malcolm calmed down, and had the perfect alibi. Both guards declared he had never left his chamber. Luke thought this was a convenient time to put Cromwell's plan into action. He asked everybody in the room to leave which made Malcolm a little uneasy, half suspecting some English military brutality. 'Sir Malcolm, I am aware of your last mission for the Scottish Government. Why did you fail to complete it?'

'I was kidnapped, and prevented from reporting to my masters.'

'You had plenty of opportunity during your captivity to get a message to Stirling. You could have asked Major Burns who represented the Scottish Government here to at least get a general report to the Government. You deliberately withheld the information to embarrass Barr, or to apply pressure on others. Or are you a covert Royalist? If you were successful in raising Dutch assistance for the Kirk controlled government it would not need to rely on the King.'

'A little far fetched Colonel.'

'Not at all. My recent trip to Edinburgh confirmed my belief that the existing government has collapsed. Scotland has two choices. Throw itself solidly behind the King, or do a deal with the English. You Sir Malcolm have to make such a choice. Or have you already made it?' Luke continued, 'General Cromwell had very fine words to say about your diplomatic endeavours over the years. He remembers you were part of the Earl of Barr's entourage in England in 1643 to arrange the alliance between ourselves against the late King. Cromwell's agents in the Netherlands have uncovered the details of your dealings with the Dutch which if made public could be embarrassing for several Scottish political leaders.'

'What are you suggesting?'

'The English will release the details of your mission and spread false rumours that you willingly gave them to us, and are in fact in hiding from your former masters. You will be seen as a traitor.'

'And how do I to prevent this?'

'Join the English. General Cromwell offers you a position in the Scottish government that he will set up after he completes his campaign against the new Royalist dominated administration.'

'Will I be protected against retaliation?'

'It won't be necessary. Already most Lowland nobles and lairds whose properties are under English occupation have co-operated without any backlash. It is the barbaric Highlanders egged on by English ultra-Royalists and Irish Catholic mercenaries that seek reprisals. Scottish landowners are sensible, and already see the virtues of the order that, at least in the short term, English military government provides.'

'Ideally I wish that the existing government with its dedication to the Kirk was continued after a number of its leading members were removed. But I am a realist. Your assessment is correct. If I were a Highlander I would throw my weight behind the King but as a Lowland landowner, and strong supporter of the Kirk an English overlordship is not too unpalatable. Stop all this nonsense Colonel; I do not have a choice. If I refuse to cooperate you tell lies and I am seen as a traitor. If I do assist you I am a traitor.' Malcolm thought for some time and then asked, 'What happens if I agree?'

'You will be sent to Edinburgh, outwardly as a prisoner of the English. You would live well in Edinburgh Castle until the time is right to reveal you as a leading member of the new pro-English government.'

'There is a one condition of my compliance. Lady Fenella Clarke must accompany me to Edinburgh.'

Luke smiled, 'You old devil. Fenella is an attractive mature woman who will never see her husband again. I can see no reason why she cannot go with you—if that is her wish. Now tell me about the Black Thistle.'

'I organized and paid for the kidnapping of Lady Elspeth, and Mistress Janet Hudson. I also suggested Mr Mackelvie be taken but it was Duff Mackail who cajoled him into visiting Castle Clarke. These abductions were part of a personal vendetta which I could afford after I inherited considerable property from a distant relative. It was personal, and revenge for a family slight.'

'So you were not responsible for Morag, Duncan, Mungo nor Sir Alistair.'

'No, their arrival made me aware that my organization had been infiltrated and someone else was using it for his own purposes. Initially I thought it was my accomplices, Derek, and Duff who had seen the potential of the situation to make money, but since their departure Mungo and Sir Alistair appear more likely the causes of the troubles we have experienced.'

'Did you order the deaths of Duncan and Aiden?'

'No, I organised the abduction of the two women I mentioned to put pressure on the Earl of Barr and indirectly assist the King's cause. I had no plans to kill anybody. Someone else implemented that agenda.' Luke brought the discussion to an end.

He immediately went to see Lady Fenella to discuss Malcolm's proposition. She would eventually be delighted to join Malcolm, but for the moment she wanted to visit her husband for a last time.

'I also wish to stay on to assist Elspeth during her childbirth. Above all Colonel, I am a woman with constant needs which you ably satisfy.'

With that Fenella threw her arms around Luke and thrust her tongue into his mouth. Somehow in their activity the couple found their way onto the only fur rug that covered the stone cobbles of the chamber. So entranced were they with each other's sexual passion and activities that they missed the evening meal, which alerted the other diners to their continuing relationship.

27

As the build up and deployment of English forces in the Western Highlands was now Luke's responsibility, he moved back to Castle Clarke. Fenella went with him to see her imprisoned husband for the last time and to continue her relationship with Luke. Luke's officers were dispersed across the region. Lieutenant Lloyd would supervise the cavalry units spread across several glens with one company directly under his command. This company was designated to shadow Sir James Cameron wherever he went. John Halliwell's company was to protect the more easterly glens against any Royalist intrusion from Stirling. Andrew remained at Greytower with a small troop. Luke spent several weeks at Castle Clarke collating intelligence regarding the movements of Sir James and any Irish incursions. The navy was convinced that the Spanish ships with the Irish mercenaries were due at any time, and the Generals-At-Sea had at least a dozen warships off the western coast. Luke, having set the forces of the English Republic in place to prosecute any action against the Royalists, returned alone on a short visit to Greytower to pursue his unfinished business.

Andrew immediately reported, 'I am worried about Lady Elspeth. She should have produced a child by now, and I would swear that her girth is less than it was a week ago.'

'I'll talk to Elspeth,' Luke replied.

'Sir Alistair is recovering. There was no infection, and he has been moving about freely. He should be sent to Edinburgh for incarceration for his own and England's safety. With conflict due, we should not be babysitting an avowed Royalist.'

'We still have a murderer to bring to justice.'

'Rubbish! The situation has changed. Ritchie, Hudson, Mackail, Clarke, Petrie and Lady Fenella have left. Mackelvie and Caddell are dead. Sir Alistair, Lady Elspeth, and Mr Macdonald are our only guests. Escort Sir Alistair and his wife to Edinburgh Castle, and execute Macdonald for the murders of Caddell and Mackelvie.'

'But we have no proof,' exclaimed Luke.

'If this were Ireland we would have proof enough,' replied Andrew.

'The execution of a Scots cleric is the last thing General Cromwell would want. He is desperate to win the Lowland Scots over to our side.'

'Then Macdonald has an accident,' concluded Andrew.

'If need be,' conceded Luke.

He visited Elspeth. The padding in her skirt had slipped to a ludicrous position. 'Elspeth, you must tell Alistair the truth. My officers are commenting on your appearance, and on the non delivery of a child.'

'Yes, now that Alistair has partially recovered he is also taking a greater interest in my appearance. I cannot hide my real condition much longer.' Uncomfortable with the deception of her husband Elspeth changed the topic and asked Luke, 'How is baby James?'

'I don't know. I have been to Castle Clarke not Edinburgh.'

'I have not spoken to you since your last visit to Edinburgh. Is my baby well? Is the wet nurse feeding him properly? Can Lady Campbell cope?' A multitude of questions flooded out of a very emotional Elspeth.

'Your baby is fine,' responded a Luke, attempting to soothe the troubled mother Lady Elspeth stared vacantly into space. Luke made no attempt to engage in conversation and waited patiently. At last Elspeth spoke,

'I will tell Alistair tonight.'

'Tell him exactly what?' queried Luke.

'The truth.'

'Is that wise? Until you are sure who wants to control the future of your child can you trust your husband with the truth?'

'What are you suggesting?'

'Until you can trust him fully, tell him you had a miscarriage. By the way where is Alistair?'

'He wanted some fresh air. Mr Macdonald and he went to the bottom level of Greytower to walk outside.'

Luke was alarmed. He left Elspeth and went to the guards outside Mungo's chamber. 'Where is Mr Macdonald?'

'He's walking with Sir Alistair.'

'Why did you not go with him? Guarding an empty room is hardly a useful exercise,' commented an annoyed Luke.

'One of Sir Alistair's guards went with them. We thought one was enough.'

Luke breathed a sigh of relief but decided nevertheless to visit the lower levels. He reached the exit door which had not been barred, and went outside. There was no one there. Luke moved to the cave. It was empty. He went through it and came out on the path that descended into a narrow glen. The path was obscured in places by the overhanging cliffs, but in a part that was visible Luke could see one of his men sitting on a rock smoking a pipe. Between the soldier and his own position he saw two men heading back up the slope. It was eerily quiet. Luke sat on one of the many rocks and waited for Alistair and Mungo to reach him. Suddenly there was long echoing scream. Luke saw what appeared to be a body hurtling towards the bottom of the cliff. He ran down the slope with sword drawn. He reached a single man peering over the edge. It was Alistair.

'What happened?' asked Luke sternly.

'Mungo was a little careless, slipped on a loose rock and slid over the edge.'

'Is that the truth?'

'No, without warning Mungo set upon me and tried to push me over the edge. He deliberately pressed against the partly healed wound on my chest. The immense pain of this must have galvanized me into a display of strength I never knew I had. As I was about to go over the precipice I grabbed Mungo and flung him away from me—and unfortunately over the edge.'

'So Mungo died as result of you defending yourself from his assault?'

'Yes, but he deserved it. It was he that shot me a month or so ago.'

'How do you know? I could never prove that it was him,' admitted Luke.

'I wasn't sure until we walked down the staircase together.'

'How did that help?'

'When the would-be killer came into the chamber and shot me, I sensed a strong smell just before I passed out. Walking down the stairs I noticed the same smell. It was Mungo's breath. It reeked of aniseed.'

'I have seen him on many occasions chewing on fennel seed, probably to cover his odorous breath. But why would he want to kill you?'

'He was acting for my father-in-law. I implied that the Earl of Barr sent me here to protect Elspeth and our coming child. That was a lie. The King sent me here to protect mother and child. Barr sent Mungo here, maybe to protect Elspeth and the child, but also to kill me.'

'Surely a man of God would not be a hired killer?'

'I thought I had seen Mungo before. Then it came to me. Several years ago he was one of Barr's clergy. He held a parish church through the Earl's patronage. He allegedly raped a young parishioner but the girl withdrew her charge. The Kirk defrocked him nevertheless for general immoral conduct, but the Earl took advantage of the man's education and appointed him as one of his secretaries. He is the man used by the Earl for many a shady activity. Now what are you going to do with me?'

'Regarding this incident, absolutely nothing. My sergeant would probably congratulate you. You may have acted in self-defence, or you may have simply murdered Mungo in cold blood. I can prove neither. I have two options. To push you over the cliff to stop whatever activity you are really up to, or take you—a committed Royalist in what will soon be a war zone—as a prisoner to Edinburgh Castle. Do you think that Mungo murdered Duncan and Aiden?'

'No, I don't think so. Barr was Aiden's marriage broker, and Duncan was a great admirer of the Earl. Both of them belonged to Barr's network.'

That evening a very small group gathered for dinner—Luke, Andrew, Alistair and Elspeth. Luke looked at Elspeth and raised his eyebrows. Her response indicated she had not yet told Alistair what had happened to the baby. The men retired to a smaller chamber and began an evening of drinking in front of a blazing fire. Gillian Shaw appointed herself as personal valet to Luke, and her intentions for later in the evening became very clear as she leant over Luke to replenish his drink, with her ample breasts almost falling out of her bodice. With the decline in guests Andrew and Luke had separate chambers, and the latter had no desire to discourage the sensuous wench.

Late into the night the three men retired all well in their cups. Luke made his way cautiously back to his room and as he carried a candle and from it lit those in his chamber he sensed someone in his room. There she was, completely naked with her hair unfurled sitting up in his bed stimulating her own nipples. Luke was surprised that she was naked, not advisable in such cold climes. The couple explored each other and tumbled repeatedly into frantic thrusting and screaming. Gillian was very noisy. Suddenly in a moment of quietness the two of them heard a piercing scream. It was Elspeth.

Luke half dressed, grabbed his sword, and raced to the sound of the screaming. Luke was first at the door and forced his way in followed quickly by three or four of his men who had been on general guard duties in the corridors. Cowed in a corner of the room was a half naked Elspeth trying to protect herself from a drunken Alistair who was laying into her with what appeared to be a leg of a chair. When he saw Luke, Alistair turned on him accusing him of being an English cur who had persuaded his wife to abort his child.

Luke ordered Alistair to be cuffed and shackled, and placed on a bed in one of the empty bedrooms. As this was being done a sedately dressed Gillian moved in to minister to Elspeth who was bleeding from the nose and mouth, and whose sensuous breast had an array of bruises. Next morning Andrew and Luke had breakfast alone, served by one of the troopers who had taken over the cooking activities. Elspeth stayed in her room with Gillian Shaw, while Alistair was taken a meagre meal by one of his guards. Later in the morning Luke visited Elspeth. Gillian did not leave the room and flagrantly pressed against Luke at every opportunity. Elspeth was too distraught to notice. 'I gather you told him of the miscarriage?' asked Luke.

'Yes, I should have waited until he was sober. But his attack on me was minor compared with what he said. He claimed that without the protection of the child my father would murder him, and remarry me to one of my own relatives to ensure that his estates stayed within the family. He shouted at me that he would thwart such activity, and had just that day killed my father's agent Mr Macdonald. He would now impregnate me immediately so that he would father another legitimate heir. Even if he and I died his family through the new baby would take all of Barr's land. He did not seem to be concerned about my present poor health. He directed all his aroused emotions into raping me. My initial screaming was a result of the pain that his animalistic assault caused me so soon after my delivery. After he ceased he smashed a chair, and came after me with one of the legs. You know the rest.'

'Gillian, could you leave us alone for a moment. There is something I need to tell Lady Elspeth in confidence.' The servant left the chamber but not before running her hand down Luke's back. Luke continued, ' You will leave immediately.'

'What will you do with Alistair?'

'Our original role here is ended. Greytower will be soon demolished. When Fenella returns she will be the only guest. You are going to Edinburgh to be with your baby, and Alistair to Castle Clarke for transport to England– and possibly to the Americas.'

28

ndrew escorted Elspeth to Edinburgh and took with him a series of letters Luke had written to Cromwell, Cobb and Sir Malcolm. After the entourage had departed Luke visited Alistair. He was immediately greeted by a tirade accusing Luke of being part of the deception. Luke lied, 'I knew nothing about it. I was surprised when I returned from Edinburgh that she had not delivered. When I left the Campbell's she was complaining about stomach pains. I later assumed it was a false alarm. Now we know the truth. No wonder she appeared ill when she returned.' Luke continued, 'I am appalled at your behaviour, Alistair. I believed you when you said that you loved your wife. How could you rape and assault her?

'One can hardly rape one's own wife, but I have no excuse for my apparent violence. I remember none of it. Can I speak to Elspeth? I am so sorry for what happened last night. I was drunk. I was completely out of control when I heard that the baby who would have consolidated our future was dead. I became deranged. And after months of abstinence I could not believe she was resisting me.'

'I am afraid Elspeth has left.'

'Not to return to her father?'

'No, she has been placed under English protection, and is on her way to Edinburgh Castle where she may be able to contact her father under her terms. I am still not sure who is the villain in this attempt to control the lost child, the Earl of Barr or yourself.'

'Will I be joining her?'

'No Alistair, you are an able bodied soldier and as our armies will soon be in conflict, I am sending you to Castle Clarke as a prisoner of war where you will probably be taken to England or with some bad luck to the West Indies.'

'But I am a wealthy laird, related to the greatest nobles in Scotland. I am the King's cousin and confidante. This is not gentlemanly behaviour.'

'In war, results not protocol or procedure, govern. There are whole islands in the Caribbean inhabited by defeated Royalist officers. But there is no hurry. Stay here until we demolish Greytower.'

That evening the dinner table was increased by two arrivals from Castle Clarke–Harry and Fenella. Fenella brought with her a number of female servants whose role at the Castle had been taken over by the English soldiers. Fenella enjoyed the assembled company. She flirted with Luke, Alistair and young Lloyd. She was amazed to hear of the death of Mungo, and Elspeth's miscarriage and her subsequent removal to Edinburgh.

Luke asked, 'When will you leave for Edinburgh to join Malcolm?'

'I am in no hurry. I brought some personal servants with me in preparation for the move. I rather like the freedom here, which I doubt will be available in that Presbyterian metropolis, or among your English Puritans. The Highlands do have some virtues.' Gillian who supervised the newcomers and was absent from the company for most of the evening nevertheless found time to make clear that she expected to visit Luke later in the evening, despite Fenella's return.

For the next few days Luke shared his attention between Fenella and Gillian. He was put out when Fenella suggested that she was ready to move to Edinburgh and asked, 'Fenella, why does a warm blooded woman like yourself want to live in the household of an elderly and cold lawyer? Sir Malcolm is a hardly an ideal husband. Or is his new found wealth a decided attraction?'

Fenella responded angrily, 'You are cruel Luke. Not every decision a woman makes is based on her sexual desires, or her need for money. I thought you would have realised by now that Sir Malcolm and I are very close.'

'How close would that be,' said Luke, with a hardly concealed snigger.

Fenella delivered a solid blow across Luke's face. 'You should have worked that out for yourself. Sir Malcolm is my father.'

Luke had once again missed the obvious, but was determined to push the new information as far as he could. 'And how did that relationship affect the initial situation at Castle Clarke?'

'I assumed that father had told you the full story when he agreed to be taken to Edinburgh although with his memory fading he can be very confused. Because of the way I was treated by the Earl of Barr regarding my betrothal, father decided to put him under pressure by asking Derek and Duff to kidnap his daughter Elspeth. He also had Janet Hudson taken so that Barr could not manipulate her future in the same way that he had done to me.'

'When did your father start to feel uncomfortable about proceedings?'

'He initially became uneasy when Alistair, Aiden, Mungo and Morag arrived. They were not part of his plan. Duff put him at his ease regarding Alistair, Aiden and Morag saying that he and Derek were taking advantage of the situation to implement a plan for the King. Subsequently I am not sure whether they were following the King or their own warped agenda, especially concerning young Morag.'

'So your father did nothing to terminate his enterprise? He was content that your husband and lover, or the King were adding agendas of their own on top of his?'

'Father was quite content until Aiden and then Duncan were murdered.'

'Why were they murdered?'

'Duff killed Duncan. Duff's sister Donna, who you met when we rescued Morag, was once the Lady Dalmabass. She was falsely accused of witchcraft by her stepson, prosecuted and tortured by Caddell, an obsessed witch hunter. He had her finger nails removed.'

'I wondered why she always wore gloves. What about Aiden?'

'I don't know for sure. A few weeks before Aiden arrived Duff told me he had met Aiden on a matter of business and that Mackelvie had proved himself a snivelling turncoat who had reneged on a sacred promise. He suggested I was better off with Derek, than I would ever have been with Aiden who would get his just desserts one way or another. Even Alistair made denigrating remarks about Aiden and had no time for him. Poor Aiden upset a lot of people.'

'And you Fenella had nothing to do with these murders? You didn't encourage Duff to rid you of two men that had contributed to your unhappiness?'

'I certainly wanted Duncan out of the way. He was beginning to annoy me with the way he was taking over the mind of young Janet, but I had less antagonism to Aiden than my father, although I found his presence at Castle Clarke a little disconcerting. I asked Derek to release Aiden, but he told me that Duff would not approve.'

'Are your father, husband and Duff, members of the Black Thistle?'

'Yes, all three. Duff said that just as General Cromwell had a group like yours, and Barr had David's unit, the King had created the Black Thistle to remove his more notorious enemies. However we women are told little.'

The inquisition ceased as Fenella appealed to Luke's instincts. The rest of the evening was lost in an alternating sequence of quiet and rowdy lovemaking. Next day Fenella and her small household were escorted to Edinburgh by a troop led by John Halliwell.

Luke returned to Castle Clarke to resume his duties. He was beginning to relish the life of a commander preparing for battle. One morning, a few weeks later, *The Providence* sailed up Loch Linnhe to the castle. Luke was astonished when he saw a senior officer, whom he recognised as Major-General Monk disembark. Monk did not waste time or words. 'Colonel, the Lord General thanks you for your work here, but demands your immediate return to Edinburgh.

I take command, as of now, of all English forces, land and sea, in the Western Highlands.'

Luke left immediately for Greytower with instructions from Monk to demolish it immediately. Cromwell's precise orders were to return with his original troop, and with Sir Alistair, to Edinburgh and to report to him as soon as possible. Cromwell had detained both Andrew and John in Edinburgh awaiting Luke's return. Luke was delayed several days as he awaited the arrival of Harry who had to return to Castle Clark to formally resign his cavalry command. Two days later two horsemen arrived at Greytower. One was Harry, the other was Donna Mackail.

Luke was apprehensive and asked, 'Why are you here? We are just about to demolish Greytower and return to Edinburgh.'

'I know. Weeks ago Fenella asked me to join her in Edinburgh where she was going to live with her father. Sir Malcolm had promised me that with a change of government he would take steps to have my conviction for witchcraft revoked and my position as the dowager Lady Dalmabass re-established. However Fenella moved before I was ready to leave. When your Lieutenant passed through my hamlet and revealed that he was heading for Edinburgh I asked if I could join him.'

The few remaining inhabitants of Greytower had an early lunch after which time was spent laying barrels of gunpowder. Major General Monk wanted no defensive positions left in the Western Highlands other than Castle Clarke. Luke felt uneasy about Donna. Her story was credible but why had she attached herself to Harry to get herself to Edinburgh? Her brother or some of his men could have escorted her. He did not trust this convicted witch. Perhaps she has bewitched young Harry.

Half way through the afternoon Luke's men, laying trails of gunpowder, had reached the upper levels of Greytower. The trails were shorter the higher the level. Luke hoped that his timings were correct and that the trooper that lit the lowest fuse had time to climb the tower and escape at the top. His men practiced their demolition strategy for the rest of the daylight hours.

Immediately after breakfast the next day the tower house was evacuated except for the four men who were to fire the gunpowder. It took several minutes for these troopers to reach the top of the tower, cross the drawbridge and reach safety. Luke had begun to feel that he had miscalculated and was just about to send one of his men back to check when he heard the first rumble followed by several explosions. Amid the dust that flew into the air Luke could see little, but he could hear an avalanche of rocks cascading down the mountainside. Greytower was no more.

An hour into the journey to Edinburgh Donna sought out Luke.

'Colonel, several days ago my brother came to me. It was if he did not expect to see me again. He told me it was time to act for the cause. The Black Thistle was on the eve of resounding victories.'

'So Duff belonged to the Black Thistle?'

'Yes, but I am sure you knew that. Behind his purely venal and criminal activities, there was a man of hidden principle.'

Luke laughed, 'And what sort of principle could that possibly be?'

'Duff like a number of young Scots resented that the King lived in England, and Scotland had in fact lost its independence. The conflict between the English Parliament and Charles I gave Scotland over a decade of autonomy. The execution of the King by the English and the divergent paths taken by Scotland and England created an opportunity to cement Scottish independence. Scotland proclaimed the young Charles as king while England opted for a republic. Duff wants young Charles to rule an independent Scotland and renounce his claims to England. He also wants the King to reverse the rigid morality and intolerance that the last ten years of Presbyterian rule has imposed on Scotland. He wants a return to the traditions and culture that flourished last century.'

'So why if this is what your brother proposes should you be in tears?'

'Cutting across this streak of principle Duff is still obsessed by our traditional feuds and vendettas. Before he dies for the King he promised me that he would complete his avenging of my misfortune.

He is going to kill my stepson, the current Laird of Dalmabass for the lies he uttered at my trial.'

'He killed the Reverend Caddell?'

'Yes. That monster received his just desserts.'

'So you want me to stop Duff murdering the laird?'

'Yes, not because I have any mercy concerning that evil man, but I would like my brother to die for his cause, and not for the murder of personal enemies. It will wipe clean the errors of his past. It will be redemption, not only for Duff but for our cause.'

29

Harry, on being told of Donna's proposal, was immediately cynical.

'Forget it, Luke. Nothing is gained saving a vicious laird from an equally vicious brigand. Let the Scots kill each other.'

'But if we help Donna we may obtain information that will help our enquiry,' countered Luke.

Harry remained unconvinced, 'Lady Dalmabass's story is flimsy. If she wanted to save her brother from himself she could have sent one of his supporters, with whom she had lived for a year or so, to warn her stepson. Those villagers would do anything for her ladyship. Or she could have gone herself. With English occupation the local authorities would not dare to intimidate her, especially as popular sentiment is on her side. How convenient it was to meet me.'

'Yes, but to what end?' asked Luke. 'She could have made her own way to Edinburgh.'

'I am not sure. It may involve Alistair. Did she deliver a message from Duff to Alistair? Maybe Duff will try to free Alistair when we weaken our escort and send some of the men to Dalmabass. Remember Luke, they are all Royalists,' proclaimed a belligerent Harry.

'If that be the case let us thwart them. You continue direct to Edinburgh with both Alistair and Donna and the entire troop. I alone will head for Dalmabass immediately it gets dark.'

Later that day Luke's plan was modified. They met a large company of English cavalry returning to Edinburgh. The heavily augmented English escort would be more than a match for any attempt by Duff to rescue the prisoners. Luke immediately headed south and as the sun rose the next morning he was at the manse of the minister he had met on his first visit. The cleric invited Luke to breakfast and asked, 'Since you were last here I gather Mr Caddell has been murdered, but I suppose you are here to investigate the local atrocity.'

'And what was that?' asked a surprised Luke.

'Three days ago a band of brigands on spotted ponies descended on the Manor of Dalmabass and murdered the laird, his wife and the three children. For good measure they slaughtered all the animals.'

'Three days ago?' muttered Luke. 'Then there is no way I could have stopped it. Her ladyship's pleading was a charade. What did she hope to achieve? Luke was silent for several minutes and then asked, 'What happens to the Dalmabass estates now?'

'If the former Lady Dalmabass's conviction for witchcraft is quashed she will, until the coming of age of more distant male relatives of the laird, administer the estates, and retain her widow's share until her death. This is a substantial amount.'

'Will her conviction be quashed?'

'I don't know. There have been lawyers out of Edinburgh demanding that the local synod reconvene to re-examine the evidence against her ladyship. I have received a number of letters from a Sir Malcolm Petrie on the matter. Unfortunately the two main protagonists in the original trial are now dead. This will make it harder to reverse the decision, as it is not customary to negate the evidence of persons who have died. If this was a plan to rehabilitate the fortunes and life of Lady Dalmabass it has not been properly thought out,' declared the minister.

'You are right reverend sir. This was not a carefully thought out act to assist Lady Dalmabass. It was a cold blooded, vicious act of revenge.' As Luke made his way to Edinburgh he wondered what Lady Dalmabass was really up to. Gillian must spy on the convicted witch.

Luke received his new orders. Cromwell had changed his mind about Barr. Whatever the political status of the man Cromwell wanted Barr under English control. Cromwell had also changed his mind about the Black Thistle. He know believed their plan to assassinate Scottish political leaders and his high command may still be active, even if the King has withdrawn his support.'

'So you want me to go to Stirling and abduct Barr? Why hasn't he been led to safety by David Burns?' asked Luke.

'That is another mystery for you to solve. Burn's deputy, Dugall Sinclair, brought us news of Barr's plight. He told us that the Earl's regiment had been incorporated into new units being formed. His senior officer David Burns has disappeared.'

'Have the Royalists executed him? It may not be politic to kill Barr but to execute his known relative and henchman might be acceptable,' mused a thoughtful Luke.

'We don't know. Find him and ascertain his current loyalties. Dugall Sinclair is still here. You will go to Stirling with him in the morning. I want Barr under English protection as soon as possible.'

Luke visited Elspeth and brought her up to date. He explained that her husband was a prisoner in the Castle; and that Fenella had joined her father Sir Malcolm Petrie, both of whom were in protective custody. Elspeth was surprised to hear that Duff's sister, the Lady Dalmabass, was staying with Fenella while the former was attempting to have her conviction for witchcraft overturned. She also was in the protective custody of the English, until these legal matters were settled.

Elspeth listened patiently but her anxiety exploded in a heartfelt question, 'And what about my future? What is happening to me?'

'I have good and bad news. The good news is that the Marquis of Argyle has agreed that you and the baby move immediately to a one of his estates.'

'And the bad?'

'The General refuses to let you go until he has consulted with your father and obtains his permission.'

'Mercy me, that is my problem. I do not want my child under the influence of my father. The Marquis would be an ideal guardian, and I would enjoy much greater personal freedom on his estates. Here or on the Barr estates I am, and would be, a prisoner. I want to leave immediately.'

'Sadly your future is outside my control. Between them the Marquis, the Lord General, and your father will make a decision. My mission involving you is over. I have been assigned to a new enterprise which will take me away for weeks. By my return your future may have been determined.'

Elspeth turned to Luke with tears in her eyes, 'Thank you for your kindness.' As Luke turned to leave she grabbed his sleeve and said, 'I wish I could repay you properly. All I can do is to warn you.'

'Warn me of what?'

'Assassination. I recall an off the cuff remark of Fenella's when I teased her regarding her continuing relationship with you. She said you were her entry to Edinburgh Castle. I thought nothing of it at the time but now I fear you have brought to Edinburgh several potential assassins: Sir Malcolm, Lady Fenella, Lady Dalmabass acting for her brother Duff Mackail, and even, for the highest of motives, my own husband, Sir Alistair.'

Luke had no time to reflect on Elspeth's advice or warning. He was soon on the road for Stirling with Dugall Sinclair. Luke hardly recognized his former ally. The Scot was strained, and had lost weight. Luke asked for details of Barr's loss of influence and the departure of David. Dugall recounted that, 'After we returned from our mission with you, the Earl revealed that both parliament and Kirk had thrown their support behind the young King to command the army. Power was slipping from his hands and his ally the Marquis of Argyle had conveniently left Stirling for his landed estates in the western lowlands. Within days the King disbanded the regiment and appointed several of the returning English and Irish royalists to positions of command in re-organised and newly formed units. When I returned to the Earl of Barr's chambers I found it guarded by strange troops and I was disarmed before entering the room. The Earl and I had a brief conversation in which I indicated that I was no longer a member of the Scots army, but was willing to continue to serve him personally. He told me that he would give me one last mission. I was to go to the English military administration, and seek refuge for him there.'

'What happened to David?'

'I have no idea. He had four options—to continue to support the Kirk party and maybe join Argyle on his estates, disappear into civilian life, join the English or become a Royalist.'

'And which do you think it is?'

'Major Burns was a pragmatic officer. He was related to Barr, but I never noticed any particular warmth between the two men, especially in recent years. Just before our joint mission he was a liaison officer at Stirling between Barr and the King. He had much opportunity to be with the King and possibly became a covert royalist. He had friends amongst the King's advisers. On the other hand our joint mission at Castle Clarke might lead outsiders to consider he was an agent or ally of the English. If he has not joined you I suspect he is with the King. He was a fanatical Scottish nationalist. A King of Scotland is a powerful icon in his vision for the future.'

Luke and Dugall reached Stirling without trouble. The grounds below the castle had been taken over by the military. The perimeter of this encampment was manned by troops who challenged those who approached. Their timing was good, as several English officers had ridden up at the same time. Luke informed the guards that they sought to serve the King. The group which Dugall and Luke had managed to join were led through the camp, told to dismount, and ordered to follow a guard to a large tent. In front of the tent there were benches on which men were sitting–eating and drinking–while they waited to enter the tent.

Dugall asked one of the drinkers, 'What's going on?'

The soldier replied, 'The new colonel-in-chief is commissioning officers to take over from the heretics that previously commanded much of our army. You can see how many volunteers we are getting. There could be more officers than men.'

Luke and Dugall finally entered the tent, but it was crowded and those at the recruiting table were obscured from those in the queue. Eventually Dugall and Luke came to a point where the queue turned and directly confronted the recruiting table. Both soldiers froze in their tracks. The Royalist colonel-in-chief was David Burns.

'I'm not surprised, but what do we do?' muttered Dugall.

'Nothing, let's see how David reacts.'

'He will have us arrested, and probably shot as spies,' bemoaned Dugall.

'Not you Dugall, you can be a genuine convert looking for a new job, just as he did. I am the English spy.'

The officer assisting David asked Dugall for his name and former military unit which he answered truthfully. David without any hint of recognition allotted Dugall to a unit, and asked an orderly to take Dugall to his new quarters. Luke gave a false name and military qualifications and received a similar destination. Dugall was almost outside the tent and Luke about to leave the recruitment table when David jumped to his feet and shouted out, 'Sergeant, a change of orders.'

Dugall and Luke dreaded what was about to follow. Dugall afterwards confessed that he expected the order to arrest the two officers and immediately execute them as traitors and spies against the King's cause. They both breathed an unexpected sigh of relief when Colonel Burns announced, 'Escort the last two officers to my own tent.' Around midday David joined them. 'Welcome gentlemen, this is an unexpected surprise.'

Luke feigned astonishment at David's new role. David was quickly on the defensive. 'Come Luke. This should not be a surprise. I have been a professional soldier for over a decade, intensely loyal to the government of Scotland, especially in maintaining its independence from England. I continue in that role. The only difference since I saw you last is that the personnel controlling the government of Scotland has changed. The Earl of Barr is out, and the King is in.'

'So why then as a loyal Scots officer opposed to England have you not revealed me as an English spy?'

'Because I need your help. I must stop those members of the Black Thistle who continue with their own agendas in defiance of the King.'

30

'**W**hy? The Black Thistle is a Royalist organization. You are now a Royalist. You should be a supporter of the Black Thistle,' muttered a cynical Luke.

'I used to be. Not only was I a member, but its second and last leader. This I cleverly concealed from you and Dugall for the whole of our joint mission. Unfortunately some members of the Black Thistle have ignored the King's orders to disband and are pursuing policies that the he does not endorse.'

'Very convenient, now the King has power he rids himself of those that helped him get there,' said Luke trying unconvincingly to take the moral high ground.

'No, it's more basic than that. When the King first came to Scotland he was driven to eliminate his rivals. He believed that it was a religious imperative to remove those English and Scots that had murdered his father. The King has now realised that to consolidate his position he must win over his current political enemies. Assassinations will only confirm his enemy's hatred. He needs all the friends that he can muster. This message has not got through to some members of the Black Thistle, or rather they are ignoring it, and embarking on an execution spree.'

'If the King is serious he can use his new found power to round up the Black Thistle and deal with them.' worst offenders are under the protection of the English. Please excuse me I must return to recruitment. Dugall you may roam about the camp, but Luke it would be politic if you stay here until I return.'

Mid afternoon Dugall returned deeply concerned. 'Luke, I don't know how far we can trust David regarding the Black Thistle. I was speaking to a former comrade who is part of the Royal bodyguard. He told me that the King has come out openly against the Black Thistle because it refuses to disband and has adopted positions completely anathema to him.'

'And what position is that?' asked Luke somewhat cynically.

'The King wishes to be restored to his English throne as well as that of Scotland, and after we defeat the English in the next month or so, the Scottish army will invade England. It will gather English Royalists as it moves south, and with the English army lying defeated here in Scotland, the King will be crowned in London, and your republic will be no more. Everything the King does in Scotland is simply a prelude to his restoration as King of England.'

'And how does that relate to the Black Thistle?'

'The Black Thistle is now a Scottish nationalist organization that is dedicated to a separate Scottish monarchy. They do not want the King to regain his English throne. He must only be King of Scotland and pursue an independent Scottish policy. If the English want the King back they should bring it about by themselves, and not by wasting Scottish resources.'

'Is this a particular worry to the King?'

'Yes, the King fears that the Black Thistle has turned its attention against many of the English Royalists now gathering in this city and against those Scots who support the invasion of England.'

'It is strange that David did not mention the King's major reason for turning on the Black Thistle. The views that are anathema to the King are pretty close to David's own position.' The discussion was brought to a sudden end by David's return.

He immediately questioned Luke, 'What are you doing masquerading as a Royalist officer?'

'One reason was to find you. Dugall was concerned. I am glad that you are well, and clearly enjoying your new role.'

'And the real reason?'

'To take the Earl of Barr to relative safety in English occupied territory.'

'Not a problem. The King will not stop it. He would encourage Barr to seek protection with the English. It will help him depict the Barr administration as treacherous, and that the King alone stands for Scottish freedom.'

'Where do we go from here? asked an anxious Dugall.

'If you are happy to serve in the new Scottish army under the King I will appoint you as my aide, and we can continue as we did under the Earl of Barr.'

'And my immediate future?' added Luke.

'We finish our old mission against the renegades of the Black Thistle.'

'I suppose that is possible with us operating in our own jurisdictions,' muttered a not completely convinced Luke.

'And I can give you the name of the founder and original leader of the Black Thistle. You already have him under lock and key in Edinburgh Castle.'

'Sir Alistair Stewart?'

'No. This man met the King on the continent just after the execution of his father. Then he was a bitter youth determined to exact revenge. This conservative diplomat said that a King could never publicly avow such things, but there were plenty of loyal supporters who would carry out such missions for him. The King accepted the proposition. The man was Sir Malcolm Petrie.'

'Good God, he had me convinced that he was a pro-English politician ready to take his place in any new English administration. Do you think his admitted feud with Barr, and the abduction of a number of persons related to Barr was separate to his leadership of the Black Thistle, or did he combine both roles in the same operation?'

'Petrie's personal vendetta initially coincided with the interests of the King but his hatred of Barr became obsessive and politically simultaneously withdrew his support from the Black Thistle. My role was to wind down its activities and re-employ its members in the legitimate service of the King. This decision was taken just before our joint mission. Aiden Mackelvie's execution may have been the last officially endorsed act of the organization. He was certainly seen by the King as a traitor. The decapitation of his brother, another Barr agent, was definitely not endorsed by the King, and reeks of Petrie's vengeance against anything connected with Barr. I took advantage of that unexpected act. Angus Mackelvie was your original contact and it was intended that he would exercise political control of our activity. I was simply to be your partner in that activity,' said David.

Luke asked, 'Given the breakdown of authority within the Black Thistle is it possible that each individual member has his own and separate agenda?'

'Yes, and they may have new victims in mind who have nothing to do with the Earl of Barr. And I know for certain that some members wish to complete the original mission of widespread assassinations.'

'How are you going to handle this current situation? Harbouring an English spy will not win you friends.'

'I am not harbouring an English spy. You have been sent here as an emissary from General Cromwell. You still have that stack of authorities allowing you to do anything in his name. You have been sent on two matters—the relocation of the Earl of Barr into English territory; and the mutual pursuit of the Black Thistle. I shall report all of this immediately to the King. Dugall will take you back to my private rooms in the Castle where you can wait for me in comfort.' On David's return the three soldiers settled down to an evening of eating and drinking. The following morning Dugall, with a troop from one of David's new regiments, accompanied Luke and the Earl of Barr towards Edinburgh. The Earl was difficult. He did not willingly accompany the new Royalist unit, convinced that Dugall had betrayed him. Barr was paranoid. He would trust nobody—until Luke produced Cromwell's authority. Luke emphasised that he had been sent by Cromwell personally to escort the Earl to Edinburgh, and the Lord General would brook no opposition. The Earl could not calm down. He expected to be assassinated at any moment. He could not accept that David and Dugall, his faithful military servants, had deserted to the King, although even Luke explained to him that professional soldiers were servants of the government. The government may have changed, but soldiers maintained their loyalty to the Scottish administration.

Barr was beginning to relax with Luke.' That may be true of Captain Sinclair but I am afraid that David Burns changed sides long before the King became dominant.'

'Did you know of his conversion to the Royalist cause before you sent him on the joint mission with us?'

'Yes, my intelligence informed me that he was highly regarded by the King, and I should no longer trust him. I could think of no better monitor of his behaviour than an elite English republican military unit. In addition I asked my civilian agent in the Western Highlands, Angus Mackelvie who had been negotiating with the Western Army, to join your enterprise as initial contact and to monitor Burns. His murder convinced me of Burns's betrayal.'

'On an entirely different issue my lord, did you order the assassination of Sir Alistair Stewart?'

Barr without warning exploded, 'An impertinent question! I will report you to General Cromwell.'

Had Luke hit a raw nerve? Was this blustering outburst a cover for Barr's guilt or a symptom of his overall distraught state? For the rest of the journey Barr refused to speak.

Three days later a sobbing Elspeth visited Luke. Behind the tears Luke discerned a cold fury as she explained her predicament.

'Yesterday General Cromwell called me to his chamber where my father was already present. Your General told him that he was a grandfather, and the child was here in the castle. Father berated me for not telling him, and ordered me to accompany him to our family castle in a few days. I refused, and asked the General if I could see my husband. Father became enraged, and verbally attacked Alistair with all sorts of outrageous accusations. In response I asked him if he was behind the attempt to kill Alistair. He did not deny it, and angrily proclaimed that he would never have his grandson brought up by a Stewart. I was sent back to my room.'

'What do you want me to do?'

'I want to tell Alistair about the baby before he hears it from someone else. And I do not want to go with father.'

Luke turned to the orderly, 'Bring Sir Alistair Stewart here!' Half an hour later two soldiers arrived with Alistair who entered Elspeth's chamber and gave her a great hug. She spoke to him quietly for a few minutes and then signalled Lady Campbell to go to the adjoining room. She returned with the baby. Alistair was pleasantly shocked. He berated his wife for lying to him, but calmed down when she argued she had done it to protect the child from her father, who was at that moment attempting to gain control of baby James. Alistair left his wife's chamber with a large smile. Luke tried to see the General that day–but had to wait almost a week. Cromwell was in no hurry to hear a subordinate berate him for forcing Elspeth to confront her father, and for revealing that she had a child.

Indeed Cromwell was not pleased. 'Luke I hope you have not developed an unsuitable relationship with Lady Elspeth. Fathers have rights over their daughters and we are not in Scotland to overturn centuries of tradition. You forget that the Earl of Barr is still a very powerful Scottish noble. Even though at the moment he is out of favour it will take only one major victory by us against the King's government, and the Scots will be pleading for Barr to return to office. I will not come between a father and a daughter. She will return to the family estates with her father the day after next.'

'General if you believe in the rights of fathers over their children and the requirement that children be with their fathers then I suggest Barr's grandson should stay here with his father. Sir Alistair has more rights than the Earl over Elspeth's child. Elspeth as a mother would be expected to stay with her child and husband, rather than with her father.'

'Enough Tremayne the political situation has determined my position. Mother and child go with Barr, and that's final. Sir Alistair is a Royalist spy.'

'He is no spy. He is an openly Royalist combatant. He has never hidden that fact.'

'Forget the Earl, Lady Elspeth and her child. These are domestic family issues without political repercussions for us. Concentrate on thwarting the remnants of the Black Thistle.'

31

An angry Luke was dismissed. He called Harry, Andrew and John to his chamber. Two days later the castle was in uproar. The Earl of Barr was beside himself with anger. Lady Elspeth and her son had disappeared. Cromwell was incandescent with rage and he knew whom to blame. He summoned Luke and delivered a blistering tirade that concluded with, 'Colonel Tremayne, you have gone behind my back and embarrassed me. I promised the Earl of Barr that his daughter and grandson would go with him when he left for his estates. You have betrayed my trust.'

'No, General, I have acted according to my word as a gentleman. Lady Elspeth was not a prisoner. She came here willingly, to give herself time to decide whether her loyalty lay with her father, or with her husband. She has now made a decision that it is not with her father, and yesterday left for parts unknown.'

'Not good enough, Tremayne. You will find Lady Elspeth, and bring her back here—as a prisoner.'

'Sir, why are we interfering in a Scottish domestic matter? The Earl of Barr would imprison his daughter on his estates and take control of the upbringing of his grandson. You are denying his son- in-law his rights as a father, which I am sure Scottish law, and the moral teachings of the Kirk endorse. I have informed the Marquis of Argyle that Lady Elspeth has left the protection of the English, but has no desire to go with her father. Until the situation of her husband clarifies she accepts Argyle's offer of protection. Argyle is a much more powerful figure than Barr. It is not in the interests of the English republic to alienate the Campbells.'

'Colonel enough! Until an agent of Argyle approaches me formally I want Lady Elspeth Stewart and her child safely within these walls. That is an order.'

Luke set off reluctantly the following morning with Andrew and six troopers for the estate of Sir Alan Campbell where he assumed Lady Elspeth would make her first stop, and Lady Campbell and the wet nurse would return home. Luke was furious that Cromwell had taken the Earl of Barr's side. His mood was not improved after being greeted coldly by Sir Alan who appeared troubled by their presence. He asked,' What brings you here?'

'We have come to fetch Lady Elspeth.'

'I do not understand Luke. You have Lady Elspeth, my wife and one of my servants under protective custody in Edinburgh Castle.'

'We had, until my General decreed that Elspeth must go with her father. This she refused to do, and with some assistance from us escaped the city to find solace here.'

'She never arrived,' said Sir Alan rather nervously.

Luke's demeanour worsened, 'My God! Elspeth and the child have been kidnapped.'

'And my wife,' muttered Sir Alan. And more audibly, 'Having released Elspeth why have you come after her?'

'The General wants her back. I am trying to buy time. Your clan chief the Marquis will protect her and I have sent one of my sergeants westward to organize the details.'

'A pretty useless enterprise if she has been abducted for a second time. You let them leave Edinburgh without an escort?' Sir Alan angrily asked.

'Lady Elspeth employed a number of male servants. A small civilian party would draw less attention, and her male servants could provide sufficient protection against the odd cutthroat. No one knew she was on the road, so her abduction could not have been planned,' countered Luke.

Luke farewelled Sir Alan. After they had left the laird's estates Andrew quietly announced, 'The man lies.'

'Evidence?'

'When I took my horse around to the stables to give it a drink one of the stable boys passed me leading a chestnut horse with a peculiar triangular white mark on its rump.'

'So what?'

'You may not have noticed, but the horse that the baby's wet nurse rode to Edinburgh had such a marking. I saw the horse in Edinburgh Castle and asked about its ownership.'

'It proves nothing. The horse may have been returned to Campbell weeks ago.'

'Maybe, but while I was near the stables I heard a baby crying.'

'Again Alan Campbell has many servants. The child could belong to any one of them.'

Luke's party rode on in silence. Luke suddenly said, 'Andrew, you may be right. After dark you and I will return and search the Campbell estates. The rest of the troop will return with the men to Edinburgh.' The night was bitter. On reaching the edge of their destination Luke and Andrew tethered their horses and proceeded on foot. The manor house was shut tight with its main doors held fast by large oaken bars. Luke and Andrew spent the night in the stable, avoiding the ostlers, who slept in one of the stalls. At dawn they waited until all these boys, but one, had left the stable. Andrew came up behind the tardy stable boy, and placed a knife across his throat.

Luke spoke quietly. 'We are not here to harm you, or your master. Your friend just led out a chestnut horse with a white triangle on its rump. Have you had that horse here for months?'

The boy who was strangely relaxed pushed Andrew's knife gently aside and spoke to Luke. 'You are the English colonel who rides the large all black horse?'

'Yes, it's a Friesian. They come from the Low Countries. I obtained mine in Ireland. Now the horse with the white triangle?'

It's been here for years, but recently it has been in Edinburgh with that gentlewoman who gave birth here. It was returned a day or so ago.'

'So the gentlewoman, your mistress and the servant returned here?'

'Yes.'

'Where are they now?'

'The gentlewoman and her men servants moved on. My mistress, the wet nurse and baby are in the big house.'

'The lady left her baby here?'

'Yes.'

'Do you know where the lady went?'

'No idea! The master does not confide in stableboys,' said the boy rather cheekily.

Luke dismissed him with a small coin and a clip on the head, and with Andrew moved to the big house. They knocked on the front door–making a lot of noise. A servant opened it, and behind him stood the ashen-faced Sir Alan. He managed a weak smile, and invited the soldiers to take breakfast with him in a small antechamber.

'I know why you have returned. I am not a very good liar, especially to people who trust me. I am sorry Colonel that I lied to you. I gave my word to Lady Elspeth that I would delay anybody who followed her as long as possible. She expected her father to send some of his henchmen to fetch her back. She spoke very kindly of your help, but did not anticipate that you would come after her.'

'Where is she, Sir Alan?' asked Luke.

'I don't know where she is, but I know where she was going.' Luke waited for what seemed an eternity for Sir Alan to reply.

'She is going to Stirling to see the King.'

'Why would she wish to see the King?' asked Luke.

'He now rules as well as reigns,' answered Sir Alan.

'I realize that, but what precisely would she want the new power in the land to do for her?' Luke responded.

'Perhaps she is going to obtain something for her father.'

'I doubt it. Her attitude to her father has hardened since he reached Edinburgh. She doesn't want him near the baby. She is reconciled to her husband. She is more likely seeking help for Sir Alistair.'

Sir Alan's face drained and he placed his head in his hands, 'Pray God that I am wrong. If Lady Elspeth has decided to throw her lot in with that of her husband so that he, and ultimately her son will inherit the Barr estates she may have fled Edinburgh anticipating an attempt to kill the Earl. And from what you told me on a earlier visit that is a Black Thistle objective.'

'Maybe the Black Thistle still has that aim, but the King has denounced assassination of his political enemies and prefers to win them over. He allowed Barr to come to Edinburgh. Once on his own estates the Earl should be perfectly safe. No Elspeth has gone to the King to effect a legal transfer of the Barr lands to her husband and son. Does the King have the necessary power to solve the problem?'

'Yes, the King could outlaw Barr, confiscate all his properties, and then allocate them to whomever he chooses. It would be logical for him to allocate the land to his distant cousin Sir Alistair Stewart and his wife. That would also forestall years of legal wrangling. Outlaws have no legal rights.'

'Despite her differences with her father I cannot see her wanting him punished to that extent. She just wants him kept away from her child. Is there any other way the King can help her?'

'The King could legally disable Sir Alistair as well as the Earl and take the lands as his own with the Lady Elspeth or rather her son as a royal ward. The King would then be entitled to the rents of the Barr estates. She would be free of both her husband and father, but could you trust this King not to exploit her?'

'That's a bit theoretical. The Barr estates are currently in an area occupied by the English military. If the rents were not paid to the legitimate landlord the English military government would take them. It would not recognise any royal claim. The King would get nothing.'

'You assume Colonel that English domination of the Lowlands is a permanent fixture. With the combination of the old Scottish army, reinforced by many English and Irish royalists, and by the Highlander armies from the north and west the English military position in Scotland is increasingly precarious. Lady Elspeth is a clever woman. She had security under English military rule, but she throws that away to seek support from the King. Her political intuition may be more correct than yours or mine.'

'You could be right Sir Alan, but Elspeth was content to stay in Edinburgh until her father arrived and undermined her sense of security. My General unfortunately threw his full support behind the Earl, and ordered Elspeth to obey her father. That is why I intervened to help her move out. I expected she would stay here until the Marquis of Argyle took her in. He has agreed, but no one has actually organized the move.'

'What are you going to do Colonel?'

'I will return to Edinburgh Castle, and inform the General.'

Luke was surprised at Cromwell's amiable demeanour on hearing that Lady Elspeth was heading for Stirling to see the King. He assumed that the General was relieved that she was no longer his responsibility. He was surprised at the General's assessment of the situation–'I know Colonel that you are cynical of our theology but you cannot deny the validity of the doctrine of providence. My victories against all odds must have convinced you that we won because God wanted us to win as part of his Divine plan. This little episode involving Lady Elspeth Stewart in its small way convinces me further that her trip north is part of the Divine plan. I am sending you back to Stirling to check out a rumour that Barr's relatives, Lord Ochilmeath and Lord Montstone, have moved there. Have they all deserted the Earl for the King? And what do they expect to achieve?'

Luke suggested his mission could best be achieved as an official emissary to the King. Cromwell agreed and Luke asked, 'And what exactly will be the formal purpose of my official mission?'

'To seek information on behalf of her distraught father about Lady Elspeth Stewart, who disappeared from English protective custody.'

On reaching Stirling Luke immediately went to the barracks of the newly reorganized regiments. His trained eye was surprised at the large decrease in the number of troops in the encampment. He was alarmed. Where were they deployed? Was there to be a surprise attack on an English garrison? When challenged, Luke asked to be taken to Captain Dugall Sinclair. Dugall was surprised to see Luke and slightly concerned. 'These meetings are dangerous as our two sides prepare for conflict. I should blindfold you if you move about the camp.'

'Pretty useless. There are dozens of English agents in Stirling, as I am sure there are Scots agents in Edinburgh Castle. Security does not exist.'

'So be it. Why are you here?'

32

'To find Lady Elspeth Stewart.'

'I thought she was under your protective custody.'

'She was. I helped her leave, but she did not go where she indicated. Instead she came to see the King.'

'Why would the daughter of the Earl of Barr leave the safety of Edinburgh to visit her father's major opponent?' asked a genuinely confused Dugall.

'You might get a different answer if you ask why would the wife of Sir Alistair Stewart want to see the now powerful head of his family?'

'Or I might ask why is any of this so important that Cromwell's leading agent wants an answer. No, Luke I do not doubt that your curiosity demands an answer to this problem, but your Lord General would not have sent you into enemy territory to satisfy your personal curiosity. Why has Cromwell sent you?

'He is concerned with the growing concentration of the Barr family in Stirling and what political implications it might have given that the Earl is under his protection.'

Dugall reported to Colonel David Burns who arranged an appointment for Luke and himself with the King. They entered the council room and found a group of men seated around a large table. The King signalled Luke and David to take the vacant seats at the end of the table– the furthest from him. Luke's prejudice against this young man was waning. The new King was like hundreds of young cornets he had trained. The young man was not full of himself–or stupid–as English republican propaganda constantly emphasized. The King spoke directly to Luke, 'Colonel, your visit is timely.

The Black Thistle has just struck. This morning two of my leading advisers were murdered. They were riding in the woods and were set upon by a gang of men riding white spotted ponies.'

'But sir, does not the Black Thistle pursue your policies?' asked Luke deliberately pretending ignorance.

'Not any longer. Sir Malcolm Petrie told me when I was on the continent that I would need a group of dedicated supporters who would carry out operations that I might not wish to acknowledge. I asked Petrie to organize such a group. When I came to Scotland I met with some of the members not far from here. Their original mission was to eliminate certain Scottish nobles, and some of your generals–anybody who had been associated with the murder of my father. Petrie seemed obsessed with a personal agenda so I appointed your companion the new leader of the Black Thistle. He was to tie up a few loose ends and dismantle the group. In fact I sent an additional agent Sir Alistair Stewart to Castle Clarke to hasten the organization's demise.'

'And now the Black Thistle has its own agenda and have expanded their assassination list to include English or Scottish Royalists?'

'Yes, they have become a group of fanatical Scottish nationalists who want to reverse the Union of the Crowns. They want me to forget my English rights, and rule simply as King of Scotland.'

'When you met these men originally did you notice anything that would be helpful for their identification?'

The King responded, 'Identification is no longer a problem. I now know who they are. Sir Derek Clarke's allies were Captain Duff Mackail, the man with the flexible wrist, and Aiden Mackelvie, with the high-pitched voice. You have Clarke under arrest, and Mackelvie was murdered. Mackail currently leads the maverick Black Thistle in its new direction but I also suspect that Petrie and his relatives may still be active wrecking vengeance on their family enemies.'

'If that be true could you issue Petrie with another order to cease any activity related to the Black Thistle and his personal revenge,' asked Luke.

'I have sent such an order several times and reinforced it by the personal intervention of two of my most trusted agents. Seeing I sacked him months ago any intervention on my part would be useless. For the moment I want you to accompany Colonel Burns and investigate the latest attack on my

English nobles. Their English servants will readily respond to you, Colonel Tremayne, than to a Scots inquisitor.'

The situation was surreal. Luke, following a request from the so-called King of Scotland, was conducting a joint investigation with his former colleague, but now potential battlefield enemy, into the murder of two English aristocrats. Luke nevertheless felt comfortable with the situation. General Cromwell had never seen the Scots as enemies, and had happy memories of his comradeship with Scottish Presbyterian officers during their campaigns against the old King. His son was pragmatic, and unlike his father, not a man of unbending principle. If the same organization was as intent on killing English royalists as well as English republicans, a temporary alliance seemed reasonable.

The only qualm that Luke experienced was that if the Black Thistle was now trying to stop Charles invading England, and thereby reducing the threat to the English Republic their activities must surely assist the English republican cause. Perhaps Luke should talk to the Black Thistle? If the opportunity arose he would. As the retinue of the murdered aristocrats were English, Luke led the questioning. It was quickly established that the nobles had taken themselves to the forest to hunt. Luke asked for a detailed account of what happened. One of the servants replied, 'I was with my master when his colleague who was some distance off, screamed. My master rode to the spot and we followed on foot. My master's friend had fallen from his horse and was like a pincushion. There were several arrows imbedded in his chest and back. Before I could warn the master to flee, a hail of arrows flew through the air and my master suffered the same fate as his friend.'

'And how did you escape?'

'It was not a problem. The assassins initially ignored us.'

'Did you see them clearly?'

'Yes, they approached us as they left.'

'Would you recognize them?'

'No, they were hooded.'

'How many?'

'At least seven or eight horsemen, their leader gave us a message for all English royalists.'

'And that was?'

'Stop the King invading England. He is King of Scotland and must recreate an independent Scottish Kingdom. English interests must never again threaten Scottish independence.'

'Was there anything about the attackers that might identify them?'

'Sir, they rode ridiculous horses,' answered another servant.

'What do you mean lad, ridiculous horses?'

'They were a lot smaller than the cavalry horses and even of the riding horses of my master, but much bigger than the New Forest ponies that I knew in my youth.'

'That helps a lot lad.'

'But they were different in another way. Although of slightly different basic colour they were all spotted. They had white spots on their rump.'

'Great!' said David who turned to Luke and nodded, 'This confirms that the assassination was the work of Duff Mackail. His men ride spotted horses.'

David reported to the King, and later returned to his tent where Luke awaited him. 'Luke, your return is delayed. This is a letter from the King commanding Sir Malcolm Petrie to give you all the information he has on The Black Thistle and for him to cease all political activity. His Majesty wishes to speak to you in private regarding Lady Elspeth Stewart before you leave.'

'So Elspeth has made it here safely. Do you know why she is here?'

'It is to do with her husband. He has been one of the King's men for a long time.'

'And a member of the Black Thistle?'

'Initially, but when the King changed his mind he sent Alistair to Castle Clarke to tell Petrie or his successor, myself, to call off the campaign of assassination.'

The King received Luke the following day. Lady Elspeth was already present. The young King spoke, 'You know my distant kinswoman whom you were very kind to during her stay at those western fortresses, and then in protective custody in Edinburgh. Thank you for assisting her escape from the demands of her father. She came here for two reasons. One is for me to protect the estates that should flow through her husband to her infant son. I hear the Earl of Barr is taking legal steps to prevent this.'

'Any legal decision one way or another will be difficult to enforce. Your decisions will be ignored in lands under English military control. Barr will not be able to enforce any legal decisions in his favour without either English or Royalist support whatever the court decides,' commented Luke.

'My grandfather dealt with extreme cases of trouble by outlawing individuals and whole clans, whose names it is forbidden to utter to this day. The thought that I might outlaw the Earl of Barr has had immediate consequences. The Earl was allowed to move into English protection as he headed home to his estates. Now the rest of his family led by his brother and brother-in- law have arrived in Stirling. They have petitioned me to keep the Barr estates within the family on the grounds that the wider family will not accept them moving into the control of a cadet Stewart branch.'

'But that neglects entirely the rights of the child's father, Sir Alistair Stewart.'

'Precisely, Sir Alistair by the laws of Scotland and the traditions of the land inherits all the lands of the Earl of Barr on the Earl's death which in turn he will pass on to his son. That is why some of the more fanatical supporters of Barr have attempted to have Alistair killed. This brings me to the second reason for Elspeth's visit. She fears that an attempt will be made on her husband's life by the supporters of her father.'

'So what are you going to do in this matter, sir?' said the republican Luke, loath to use any titles or honorific the young King might expect.

'Absolutely nothing,' was the unexpected reply. 'I cannot give Lady Elspeth real protection. As you know I have no formal court here. This is a military encampment and we are soon to move against you. Lady Elspeth, despite your support, is not safe in Edinburgh, as your General seems to take her father's side in this dispute. She believes that she will be safe under the protection of the Marquis of Argyle who remains one of my senior ministers, although he is currently absent on his estates. When the situation clarifies and my authority is established throughout Scotland I will be able to offer her the protection she craves. In the short term I need your help to protect Sir Alistair and his wife. When I reject the requests of the Barr claimants to give them the family lands it will put pressure on the disappointed to remove Sir Alistair, and kidnap the child.'

'May Lady Elspeth return with me to Edinburgh? She can see her husband, and then I will personally escort her and the child to the Marquis of Argyle. My General will not oppose the decision of the Marquis to offer her protection. He is careful not to alienate the great nobles of Scotland.'

The King nodded. Lady Elspeth backed out of the room. Luke did not have the temerity to turn his back on the King. He sidled out with his body side on to that of the would- be monarch.

The next day Luke and Elspeth left for Edinburgh. Only Captain Cobb and Sir Alistair were informed of their arrival. Luke released Alistair from his confinement, and invited him to share his own chamber, explaining that the King's decision regarding the Barr estates would make Alistair's demise an imperative for the Barr family. Luke took the opportunity of their new camaraderie to question Alistair about his role at Castle Clarke and Greytower. Alistair explained, 'I was a founding member of the Black Thistle, and apart from Sir Malcolm Petrie the only one known to the King for some time. When the King decided his policy was mistaken he sent me to tell Malcolm to call off his plan and release the kidnapped victims. The fact that one of these was my wife made my arrival more acceptable.'

'Did Malcolm listen to you?'

'I am not sure. He muttered that some aspects of the situation at Castle Clarke were not related to the Black Thistle but were manifestations of a private feud, aspects of which he wished to continue. He was disconcerted when I pointed out that one of the victims was in fact my wife whom the King had especially wanted released.'

'Why do you say you are not sure if he listened to you?'

'After I gave him the King's order matters continued to escalate at Castle Clarke and later at the tower–murders, attempted murders, and further abductions.'

'How did he explain what was happening?'

'He was completely confused. He claimed that he knew nothing of the murders and I feel increasingly that he wanted out. I knew that he had been replaced as the overall head of the Black Thistle, but I did not know then who it was, or whether he was at Castle Clarke.'

'Did he name the members of the Black Thistle?'

'No, like you I was not sure who were members, and who were Derek and Duff's employees. That duo carried out the initial operations for the Black Thistle in the Western Highlands, and also for Malcolm's independent family feud.'

'It's time to ruthlessly interrogate Sir Malcolm Petrie,' concluded Luke.

33

Sir Malcolm was confused. He either could not remember, or did not know. Luke's patience was quickly exhausted. 'Sir Malcolm stop this charade. Your so-called King, Charles Stuart, orders you to reveal to me any matters that may help me, and his own men, apprehend remaining members of the Black Thistle. These mavericks have turned on the King's English supporters, or continue to pursue discarded policies, such as the assassination of political leaders.'

Luke asked an orderly to fetch Alistair. Malcolm was slightly surprised by this development. Luke took advantage of the old man's discomfort. 'I have in this room two members of the Black Thistle, the clandestine organization set up by you, Sir Malcolm, to carry out the more unsavoury wishes of Charles Stuart. Your initial priority was to assassinate English and Scottish leaders who had participated in his father's execution, or who opposed his participation in the government of Scotland. When Charles saw the error of such a policy he sent Sir Alistair to disband the Black Thistle and release the prisoners you had. Why did you disobey the King's orders?'

'I did not disobey the King. Yes, Sir Alistair came to Castle Clarke claiming authority from the King, and ordering me to cease operations. The King told me when he replaced me as leader that the only authority I should accept regarding the Black Thistle was from the new leader who would be wearing the King's amethyst ring. Alistair wore no such ring.'

'So you ignored him.'

'Yes, especially as the new leader appeared shortly afterwards, flashed his amethyst ring, and told me that our original orders were simply on hold until the situation clarified. We should continue to keep ourselves in readiness to act.'

Luke probed further, 'The leader of the Black Thistle approached you at Castle Clarke and modified the orders delivered by Alistair?'

'Yes, it occurred in Castle Clarke just after your troops took over the castle. The man knocked on my door and when he entered the room he extinguished the candles. He was hooded and wore a large cape. In the pale moonlight he displayed his ring and gave me his orders.'

'Did you know who it was?'

'No, he was male, and Scots.'

'Did these conflicting instructions confuse you?'

'No, the man with the ring was the King's true representative. I thought Alistair was falsely claiming an association with the King so as I would release his wife. But other matters had me worried. In the first place Castle Clarke filled with persons I had not abducted. Morag Ritchie and Mungo Macdonald were not on my list. I had contemplated taking Cadell and Mackelvie but they arrived before I had actually authorised the kidnapping. Then Mackelvie was murdered, and later Caddell suffered the same fate. I was not responsible.'

'But you know who was?'

'No, I do not.'

'But you have your suspicions?'

The older man trembled, and Luke feared he was going to regress into his poor memory phase. 'For a time I suspected my daughter Fenella of using her charms on Duff to have Aiden murdered. He had been her betrothed, and my own personal conspiracy was to punish those who had been responsible for her humiliation. The Earl of Barr had removed her from his brother-in-law's house, and Mackelvie had readily cast her aside for one of her companions on the alleged grounds of gross immorality.'

'What about Caddell?'

'Even if Fenella organized Aiden's demise I am sure she had nothing to do with Caddell. She had no deep personal feelings against Caddell. If she had they arose during their mutual time in Castle Clarke. Fenella hated the way Caddell wormed his way into the mind of that young innocent Janet Hudson.'

'Who were the members of the Black Thistle at Castle Clarke?'

'Derek, Duff and myself–and the new leader with the amethyst ring. Despite Alistair's claims I had no proof that he was a member, or in any way acted for the King.'

'We now know by his own confession that it was David Burns who wore the amethyst ring. Did you never suspect that he was your new leader?'

'No, I always thought he was a puppet of the English despite, his professed loyalty to the Earl of Barr.'

'Who redirected the Black Thistle's political antagonism towards those Royalists who wanted to divert Charles's energies into a futile attack on England? Was it David?' asked Luke.

'I do not know. I did not know there was such a change of direction. I have had no contact with any members since Derek and Duff left Greytower.'

'Do you know which leading Royalists want the King to concentrate on an independent Scotland–and forget his English claims?'

'Ironically enough it comes back to the Earl of Barr. Over a decade Scotland has been independent of England and pursuing its own policy. Argyle and Barr saw the great advantages in this practical independence. Barr's reluctance to give the King any real power was in part driven by the fear that the King would develop a British, even an English interest against the wellbeing of Scotland.'

'The Earl of Barr's resistance to the King has led to his political overthrow. He has little political influence left.'

'But many of his family and former supporters, who are rallying to the King, retain this fervent Scots nationalism.'

'Which family members?'

'Montstone, and your former colleague David Burns always had somewhat fanatical views on the subject.'

'You used the Black Thistle in part to get back on Barr. Do you believe it has been taken over by Barr's men for their own ends?'

'Anybody could now be manipulating the Black Thistle, including the English.'

'Ridiculous!' exploded Luke.

'Is it Colonel? Who benefits most if the Scots pull out from conflict with England, and refuse to invade your country. It would be a major coup to remove the Scottish menace without a massive loss of life that battles entail.'

'You misunderstand us. We are not in Scotland to prevent you invading England. We are in Scotland to make sure that the so- called King will not be restored to the English throne. It would suit us for a pathetic Scots army to invade England. We could settle the matter once and for all.'

'Perhaps Colonel your government does not always put the military in the picture,' Sir Malcolm persisted. Luke was annoyed, and ended the interview.

He motioned for Alistair to follow him to his own chamber. Malcolm's parting shot had found its mark. Could the murder of English royalists in Scotland have been a plot organized by the government of the English Republic? The politicians did not trust the military, and had their own intelligence. He must alert the General. Luke and Alistair had several draughts of whisky, and a midday meal of boiled mutton. In the mid afternoon they received an invitation to dine that very evening with the Lady Fenella and Lady Dalmabass. On their arrival Fenella was blunt. 'Luke we have been here for weeks and you did not pay us a single visit.'

'I have been out of Edinburgh for most of that time. I made two trips to Stirling,' replied Luke defensively.

'And you Alistair, I did not know you were in the Castle until father told me today,' continued Fenella.

'Until today I was Luke's prisoner,' answered Alistair.

'So why are you now his companion?' asked Donna sharply.

'We discovered we had the same enemies,' answered Luke.

'Still hunting the ephemeral Black Thistle? Father closed it down after the murder of Mackelvie,' interposed Fenella.

'Your father may have closed it down in his mind, but it still operates. It murdered two English aristocrats last week. But you did not invite us to discuss the Black Thistle–or did you?'

'Then why did we invite you Luke?' Donna asked.

'To berate me for questioning Sir Malcolm so intently. Is he joining us for dinner?' Luke replied.

'The answer is no to both questions. Your examination wearied him so much that he had a light tea and retired to bed some time ago. And no, we do not wish to berate you, but to caution you against accepting what father has told you,' replied Fenella.

'And why should we disregard the vital information Sir Malcolm has given us?' said Luke, deliberately inflating the importance of the material offered by the old man.

'The incidents at Castle Clarke and Greytower, and subsequently here, have proved too much for him. He is old, and his mind wanders. He confuses events of decades ago with what happened yesterday,' Fenella explained.

'So any information he gave us regarding potential Black Thistle targets are memories of earlier activities, and not future objectives?' Alistair surmised.

Both men were on the same wavelength–convince Fenella and Donna that Malcolm had given them valuable information. It might lead to one of the women adding to their knowledge of the Black Thistle. Luke wondered if Donna had been in touch with her brother. He waited until they were seated at table and well into the sumptuous meal prepared by Fenella's servants. Luke turned to Donna and asked quietly, 'Have you seen Duff in recent weeks?'

Donna slowly lifted her gaze from the food on her plate and gave Luke a withering glare. 'My brother has a price on his head. Both the Highlanders under James Cameron and the English military are hunting for him high and low. He would not risk his head by coming anywhere near Edinburgh.'

'That's true enough,' said Alistair before Luke could stop him.

'Now the King also is after him. He is murdering in the vicinity of Stirling.'

'What do you mean?' asked a now anxious Donna.

'Your brother killed two English aristocrats last week.'

'Evidence sir?' demanded Donna.

Luke decided to end the discussion. 'I am sorry that information is sensitive. Let's not ruin this delightful evening by talking politics.' Fenella hurriedly agreed, 'Yes, Luke. And we have not yet reached the real reason we invited you this evening. Are you enjoying the meal?'

'It is the most delightful meal I have had in Scotland,' said the meat eating Luke, tired of the oat dominated repasts of the last few months.

'Would the Lord General and his senior officers accept an invitation to dine with us? I have the best cooks in the Castle and would like to thank the Lord General for giving us protection at this difficult time,' purred Fenella.

'I doubt whether he would accept personally. The General is a frugal man, and often unwell. Invite him, and I will put in a good word for you,' replied an immediately suspicious Luke. Fenella had overplayed her hand. The evening continued for some hours until both men, well overtaken by drink, excused themselves. Luke had felt unusually light-headed, but had steeled himself against talking too freely about matters of military and political importance. He sensed that Donna was asking sharp questions about what he and Alistair were about. Their new alliance had clearly disconcerted the women. Why?

They had not proceeded far down the corridor that led away from Fenella's apartment when Gillian Shaw overtook Luke. Alistair returned to Luke's chambers alone. Luke followed Gillian back to her bedchamber. After a few minutes of physical activity the poorly performing Luke fell asleep. When he awoke Gillian was already dressed and going about her duties. They had one final grope at each other and Gillian whispered in Luke's ear. 'Lady Dalmabass lies.

She told the mistress yesterday that she had met her brother the day before and that everything was well in hand.'

'I wonder what "well in hand" refers to? Have you overheard anything else I should know?'

'No, because of our relationship Lady Fenella does not confide in me.'

'That's a pity. I would give anything to know what brother and sister discussed. I have a bad feeling that we will soon suffer the consequences of their conversation.'

34

Luke and Andrew leaned over the castle's parapet. They were looking downtown in the direction of Holyrood House. Luke idly followed the progress of a large number of men, some walking, some on horseback, making their way from the castle down into the town. Andrew whistled and calmly said, 'There's trouble. Armed horsemen are galloping directly at those who have just left the castle.'

'My God,' said Luke. 'Look at the horses. Spotted ponies. It's Duff Mackail.' The Officer on Duty had assessed the potential situation and frantic cornet calls soon had a detachment of the castle's garrison riding post haste to assist the amblers. 'Whose retinue receives such a quick response from our garrison?' asked Andrew, amazed at the rapid reaction.

'Let's find out.'

The Officer of the Guard, Luke's old comrade Captain Cobb, watched developments below him with a mixture of concern and excitement. The English troop reached the fracas just in time to prevent wholesale slaughter. The assailants did not stay to engage the troopers, and scattered in different directions.

'Your response was remarkably quick Cobb. Who is being attacked?' asked Luke.

'The Earl of Barr. General Cromwell was not pleased, but the Earl insisted on visiting some old friends in the town. He refused an English escort. He believed that it would give the people of Edinburgh the wrong impression of his political loyalties. But I had a troop concealed just below the castle entrance–if needed.'

'It doesn't look too good down there. There are several bodies on the ground,' observed Andrew.

Cobb immediately sent several wagons down the slope to collect the dead and wounded. Luke and Andrew were soon at the scene. Luke asked the cornet in command for an assessment of the situation.

'There are two dead servants, three dead soldiers and two wounded noblemen, one seriously. The nobleman near death is the Earl. He has a javelin wound in the chest, and a pistol wound in the face. The attackers used lances which they thrust through the ineffective swords of the defenders, but one attacker was armed with a hunting javelin which he threw successfully at his lordship, and another fired almost point blank into his body.'

'Are there no dead attackers?'

'No, the Earl's party was taken by surprise, and defended themselves inappropriately with sword alone. Not very effective against light cavalry armed with lances.'

The Earl's body was placed on a wagon and taken to the garrison's hospital. Cromwell summoned his most proficient military surgeons and ordered Captain Cobb to ensure that any party leaving the castle should be accompanied by sufficient pikemen to withstand any similar cavalry attack. Luke immediately told Alistair that his father-in-law was wounded–perhaps fatally. Alistair gently told his wife of her father's condition. Luke sent John, who had just returned from negotiating with the Marquis of Argyle, to Sir Alan Campbell to collect the Stewart baby. He was to take it immediately to the Marquis. If the Earl died that baby would become a critical element in the future of Scotland.

Cromwell sent a message to Stirling to inform any of the Barr family in the town that the Earl was in a critical condition, and that they would be granted safe passage through the English lines. For a week the Earl's relatives passed in and out of his bedchamber. He was conscious and rational, though very weak. Elspeth spent hours with her father. She assured him that his grandchild was safe and under the protection of the Campbells. The Earl's physicians considered any discussion with him concerning her husband was however inappropriate. Three weeks passed and all of Barr's close relatives remained in Edinburgh, except for David Burns who stayed only long enough to ascertain first hand the real condition of his wife's uncle. Events at Stirling demanded his immediate return. Luke wondered what precisely these might be. Decorum and protocol both required such a close relative to have spent more time with a dying relation. David's rapid return to Stirling and the movement of so many troops away from that city had the English high command deeply concerned. The Earl was moved back into his own chambers in the castle and attended to by his own staff. He gradually improved.

Two weeks later Luke was urgently summoned from a leisurely game of cards to Cromwell's private apartments. The general was red faced and agitated. His customary fast pacing had almost accelerated to a run. 'Luke, we have a crisis. The Earl is dead. I have sealed off his bedchamber and want you to immediately investigate the situation.'

'What's the crisis? I have seen many a war-wounded soldier appear to recover, and then without warning relapse and die within hours.'

'Talk to his physicians. They are dismayed.'

Luke questioned the two physicians who had been looking after the Earl. They had been the Earl's doctors for over a decade. Both affirmed that he was improving on every visit and that they too had seen many victims of war who appeared to recover but then relapsed. They both denied that the Earl's case was similar. 'Why not?' asked Luke.

The senior of the doctors asked Luke to follow them into the bedroom where the body of the Earl lay. 'Colonel, look at his lordship!'

Luke was shocked. The face of the Earl was distorted. It was twisted into a hideous visage. Barr had seen some horrible image at the moment of death. He had been frightened to death. Luke asked the physicians if they had seen anything like it before. The younger doctor answered for them, 'Death by poisoning sometimes can have a similar effect, but the fear frozen in his lordship's face is something I have never experienced. But I have read of such appearances.'

'So what causes it?' asked Luke, anxious for an answer.

The doctor appeared a little nervous before replying, 'Not all medical men would accept the explanation, but victims of witchcraft have manifested similar symptoms.'

'Witchcraft!' exclaimed Luke. 'Is there any truth in such a claim?'

'If patients are fed particular herbs they might imagine that Satan was dancing at the end of the bed. These potions create hallucinations, and depending on what the patient saw he could be terrified.'

'Did you give the Earl any mind-bending drugs?'

'No, our remedies are harmless. At the best they relieve pain.'

'What caused the earl's death?'

'He was poisoned, but not by us. And we do not know if the poisons killed him directly, or they so affected his mind that he died of fright.'

Barr had feared poisoning for decades and for years employed two tasters. They tasted two meals each and one of the four was given at random to his lordship. Luke asked Barr's steward, 'If he was not poisoned by his doctors, or by his cooks was there any other way poison could have been administered to him?'

'Yes,' exclaimed the steward. 'The Earl used a personal healer all of his life. She administered potions and ointments to him whenever he felt unwell. She was adept in giving him hot herbal effusions.'

'Could she have killed him?'

'Not knowingly. She is an old family retainer whose grandmother had exercised the same role. When the Earl became a powerful political figure Mother Ross came to Edinburgh as part of his retinue. She is devoted to his lordship. And it is rumoured that she used her potions with much success against his enemies.'

'Did she see much of him during recent days?'

'Yes, she came to give him the hot potions three times a day.'

Luke sent for Mother Ross who was very distraught at the death of her master. Luke went straight to the point, 'Your master was poisoned, and we have ruled out the doctors and the cooks. The poison was in the potions you made him drink. What ingredients did you use?'

The old woman began to tremble uncontrollably, 'Sir, I did not aim to kill my master. Without his protection I would have been tried for witchcraft decades ago. I fear after this the Kirk will arrest me on such a charge.'

She led Luke into a tiny room which she had inhabited since she came to the Castle. He brought the physicians with him and was surprised that they recognized and passed as relatively harmless the ingredients that Mother Ross had used in the potions. They were basically effusions of thyme and sage. Luke then noticed an empty phial. He turned on the old woman, 'What was in this?'

'A herb that induces sleep.'

'So you used the last of it on the Earl just before he died?'

'Yes, I had run out of it before my last visit to his lordship. I asked the cook did she know where I could obtain more. The cook said she had heard that a gentlewoman within the castle had a collection of herbs, and might be able to help me.'

'Did you meet this woman?'

'Yes. She gave me some of the herb I was after. It was slightly different to what I usually crush. It was purplish green rather than greenish purple but in other respects it looked and smelt the same.'

'So if this herb was not what you thought it was, it could have killed the Earl?'

'Yes.'

'This gentlewoman, can you tell me any thing about her?'

'I do not know her name, or where she lives in the castle. The cook told her what I wanted and she brought it to me at night.'

'Is there nothing that would help me identify her?'

'Yes. She has a deformity. When she took off her gloves to separate the herbs I saw that she had no fingernails.'

Luke dropped the empty phial.

Later he told Harry who remarked, 'Brother fails, sister finishes the task. Perhaps the Scots were correct when they convicted her for witchcraft. Whatever she put into Mother Ross's potion turned his lordship's mind. What are you going to do?'

'Arrest Donna secretly, without warning, and remove her to the deepest dungeon where none of her friends will find her.' Within half and hour Donna was imprisoned in the castle's dungeon. Luke waited to see how her acquaintances reacted to her inexplicable disappearance. Their reaction astonished him. There was none. No one informed the authorities that Donna was missing. To elicit some sort of response from Fenella or Malcolm Harry put about the rumour that she had left the castle. Still no reaction. Luke forced the issue. He visited Fenella who received him warmly and explained that Sir Malcolm was asleep, and her various servants dispersed throughout the castle and town. She was alone.

Luke, with a regular liaison with Gillian, did not need the debilitating sexual exercise that Fenella had in mind. Nevertheless he participated in what for both of them was hardly a subdued carnal experience. Fenella was still not satisfied and expressed her displeasure by attacking Luke's preference for her servant–and maybe her friend Donna. This allegation created an opening for Luke who asked, 'If I was having an affair with Donna it would have been on hold for several days. Where is she?'

'You have her hidden away in a comfortable apartment in the town,' said Fenella with a tinge of jealousy.

'Don't be ridiculous. I have been open about my relationships, and it would be very difficult to conceal them in this environment. I have never shown any interest in Donna.'

'She must be the only women in Scotland you could say that about,' chuckled Fenella, who was slowly regaining her composure.

'Donna must be on some secret enterprise. What is it Fenella?' reiterated Luke.

'Donna is a free agent. She comes and goes as she pleases.'

'Did she leave any message for you?'

'No, but she is anxious to have her conviction quashed, and wanted to speak personally to several persons involved in the process.'

'Trying to influence the result of her appeal is she?'

'Yes, a convicted witch in Scotland is vulnerable to all sorts of fanatics, and unbending officials. She needs all the influence she can muster,' explained Fenella.

Luke was pleased with himself. The witch could have murdered half the castle if he had not had her locked away. His reverie on Donna's murderous fantasy came abruptly to an end. Fenella dragged him back on to the bed. It was morning when Luke left her chamber.

35

A week elapsed. Luke was invited to the reception held by Malcolm and Fenella for the senior officers of the English army, and members of the Earl of Barr's family before they moved with the body to the family seat. Luke was on edge. The Black Thistle was originally concerned with the Earl of Barr and his family. Luke knew that the Black Thistle through the Mackail's had killed the Earl on the second attempt. Why would Malcolm now be inviting his archenemies to this reception unless it was to deliver a similar fate? Secondly it had also been the aim of the Black Thistle to eliminate the English military leadership. Perhaps Fenella believed that Donna had gone off to help her brother, or in some way to prepare for this reception in which they would murder their enemies, and achieve their ends. Fenella, Malcolm and Duff intended to implement the original policy of the Black Thistle of eliminating Scottish nobles associated with the Earl of Barr, and the English military command.

Luke informed Cromwell outlining his specific fears and Elspeth's warning. The General authorised him to take counter measures. The castle was not placed on obvious alert. This would warn the conspirators that their plot was being slowly unravelled. However entry to and from the castle was carefully regulated on the grounds that those who initially attacked Barr may attempt to attack his relatives. Luke took other steps that he concealed even from most of his own men. The day of the reception arrived. Luke had thought hard about the most effective way the Black Thistle could eliminate such a large number of people. When he heard that given the size of the reception Fenella had sought permission to hold it on the ground floor of one of the recently renovated towers which not only accommodated the guests but gave easy access for the food to be brought from Fenella's own apartments, he was elated. Luke discussed the immediate situation with Andrew. 'The so-called King of Scots real great grandfather was blown away with a bomb, and the lad's grandfather nearly suffered the same fate at Westminster on the fifth of November. Let us search the under chamber of the renovated tower.'

Andrew advised, 'Let's not make it obvious. I will pretend to be a mason repairing a fallen wall.'

Andrew took over an hour searching the under croft. He emerged singing some bawdy country song, and jubilantly reported,

'Your hunch was correct. There are a dozen barrels of gunpowder all disguised as food. The top few inches of each barrel is either salt, salted fish, salted meat, dry biscuits or oatcakes. The barrels have obviously entered the castle openly as food supplies for this afternoon's reception. Stacking them in the under croft would appear logical, and cause no suspicion.'

'How are they to be ignited?'

'There are trails of gunpowder across the floor starting at the door. Someone will ignite the powder trail with a light from the fires keeping the food warm. I have thwarted this plan by sweeping away part of each trail. They can be lit but will peter out before they reach the barrels.'

'Good work Andrew. We can now enjoy the reception. And watch the conspirators fall into our trap.'

The reception was well under way when Andrew approached Luke, 'What's going on? The General doesn't look himself, and I do not recognise any of the other officers. I see they are all wearing thick protective leather doublets.'

'Yes, Fenella also noticed the doublets, so I told her it was a special jerkin which English officers wore on official occasions.'

Luke noticed that the Barr group of mourners were there. Alistair and Elspeth; David, who had returned just for the occasion, and his wife; Lord Ochilmeath; and Lord and Lady Montstone and a dozen or so more distant relatives. The group imbibed heavily. Some of the women dozed off. Andrew who had not held back began to feel faint. Guests all over the room were moving in and out of consciousness. Luke did not drink, and remained alert. As the guests tumbled over Luke also noticed other ominous developments.

The servants who had brought the meal, and attended to the guests were new. Luke recognised two of them from his days at Castle Clarke and Greytower. They were Mackail's men.

Luke almost panicked. The guests were drugged, the servants were a gang of deadly cutthroats in the employ of the Black Thistle, and the under croft of the room had been stacked with gunpowder, which luckily Andrew had rendered useless. Suddenly Luke became aware that most of the servants had left the room. Malcolm could not be seen and Fenella was about to exit following a retainer who carried a firebrand. Luke followed. He entered the under croft as the servant was about to fire the gunpowder trail under Fenella's direction.'

Luke had a primed pistol in one hand and his drawn sword in the other. 'Stop my man. If you don't I will shoot Lady Fenella, and slit your throat.'

The man jeered, 'Too late Englishman.' He dropped his firebrand onto the gunpowder trail.

'No, my man. We have swept away the mid portion of the trail. Your fire will peter out before it ignites the barrels. Come my lady, we will all leave here.' Luke escorted Fenella and the servant from the under croft and marched them into the middle of the courtyard where detachments of Cobb's men, as arranged, had the reception area surrounded. Luke informed Cobb that most of the guests were asleep having been drugged, and would need to be helped out. Andrew joined them, announcing that some guests had recovered and had been taken to their quarters by another of Cobb's detachments. The three men were about to report to the reception hall to assess the situation when there was a massive blast. Shrapnel from the tower flew through the air killing some of the men in the courtyard, and wounding others. The tower collapsed. After the clouds of dust an immense fireball emerged which drove back several of the potential rescuers. Somehow, despite Andrew's precautions, the barrels of gunpowder had exploded.

Luke looked at Fenella. Her smile reflected her delight. She gloated, 'We have won. The Barrs are obliterated and your senior officers killed. The original mission of the Black Thistle is complete. And at this very moment we will be abducting the King.'

As Cobb led Fenella away, Luke and Andrew waited with other soldiers for the thick smoke to clear. A long line of men passed buckets of water to a small group attacking the flames. As the flames were extinguished, the soldiers led by Luke removed the fallen beams and stones. Cries could be heard and a number of the victims were rescued alive. By nightfall Luke had a clear picture of the casualties. All the English officers including one who bore a remarkable likeness to Oliver Cromwell were dead. None were currently serving officers. Luke had approached Cobb to release a number of military prisoners, promising them their freedom if they impersonated England's high command, and enjoyed a reception replete with drink and food. They had not been forewarned of a possible attack on them as Luke was convinced that any attack had been thwarted.

Alistair, Elspeth, and David had left the chamber before the explosion, although David's wife had died. Ochilmeath and the Montstones and most of their retinue were also dead. Amongst the few rescued from the ruin was Gillian Shaw who only suffered a series of bruises where dislodged stones had fallen on her. Malcolm was unaccounted for. Cromwell was furious. 'If we had not guessed what might have happened the whole English high command in Scotland would have been killed. As it is we have lost three leading members of the Barr family in recent weeks, and you have lost a major conspirator in Sir Malcolm Petrie. How did the gunpowder explode? You told me that your man had prevented any possible explosion.'

'I do not know what happened but the Black Thistle is still operative. I have warned David and Alistair that an attempt will be made to abduct their King.'

Cromwell was silent for some time. Eventually he spoke again, 'The boy maybe his father's son but he is not his father. Perhaps we may be able to talk the lad into some sense. If an Englishman under orders from the Lord General saved his life, or prevented his abduction, it would be a major diplomatic triumph. Go with your Scottish friends and thwart the Black Thistle. I will give you a letter to Charles Stuart officially warning him of the plot to kidnap him.'

David, who had only arrived in Edinburgh the day before the reception, insisted he ride post haste to warn the King. Alistair prayed that David would reach the King in time. Luke and he were delighted on reaching Stirling to be assured that the King had not left his heavily guarded chamber since David's arrival, and that Dugall Sinclair commanded the royal guard. The King received Alistair warmly with a prolonged hug. 'I am glad that Lady Elspeth and yourself are safe.'

Luke proffered the King a sealed letter from the Lord General. The King perused the letter for some time, 'You are aware of the content of this epistle. Your General believes that the Black Thistle has been taken over by a group of malcontents who want to ensure that I maintain an independent Scotland and not follow up my rightful claims to the English throne. He suggests that they will not, like the nefarious English, kill their anointed King, but will kidnap me and insist I rule an independent Scotland according to their interest, an old Scottish custom. Colonel, do you agree with your General?'

'Since your father tried to change the religion of this country and was defeated by the Presbyterian led nobility, Scotland has pursued its own policy. These nobles have had a decade of Scottish independence. When you were subservient to this Scottish government they did not fear any loss of independence except through English military intervention. Unfortunately English successes against the Scots forced the former government to hand over its power to you.

Now you indicate that you will use Scottish resources to return to the English throne. To many Scots this is to subjugate Scottish interests once again to those of England. This is already manifest in your extravagant support for the large number of English and Irish gathering here in Stirling.'

'Colonel, despite your republican pretensions I am the rightful ruler of England and Ireland. I have been crowned King of Scotland but your usurping republican military have prevented me from returning to my English and Irish inheritance. I am sure that my loyal Scots subjects will rally to my cause to firstly remove your army from Scotland, and then on the back of that success invade England, and restore me to the English throne.'

'Sir, I do not disagree with that as a possible scenario, but your immediate problem is that there is a small group of Scots who welcome you as their King but do not want you to dissipate Scottish resources on a hare brained scheme to invade England– especially given the strength of our army.'

Luke's discussion with Charles Stuart was interrupted by a heated argument in the far corner of the room between David and Dugall. Alistair admonished the soldiers for raising their voices in the presence of the King. David apologised, 'I am sorry sire but given the seriousness of the situation I order Major Sinclair to stand down his men from guard duty, and replace them with fresh troops of my own. He has refused my direct order.'

The King was perplexed. He turned to Alistair, 'Cousin, escort Colonel Tremayne to a chamber reserved for him in the castle. He leaves for Edinburgh in the morning. Colonel Burns you may change the guard, and relieve Major Sinclair and his men from their duty.' David immediately responded, 'Thank you sire.' He went to the door of the chamber, and a large troop of soldiers poured into the room. Dugall's men were led from the chamber, but the King indicated that Dugall must stay with him. Alistair and Luke left the King's room with Dugall's men. Alistair did not move far from the King's door and motioned Luke towards an alcove. He whispered,

'What did you make of the argument between Dugall and David?'

'Unexpected. Dugall is a professional soldier through and through, and I have not known David to throw his weight about, especially in the presence of the King.'

'There is something that I must tell you. It begins to take on a serious aspect. At the reception at Edinburgh Castle just before the explosion killed most of those present David suggested that Elspeth and I should leave the room to take some fresh air. He almost pushed Elspeth out of the room, and was eager that we move away from the building. At the time I put it down to David's known dislike for such functions, and the presence of so many English officers in the room. Now I think David knew the bomb was about to explode, and did not wish too many innocents to die.'

'Do you think David, in defiance of the King, is still the active leader of the Black Thistle?'

'No, the Black Thistle under Mackail is out of control and following its own agenda. Some of this however coincides with David's vision for a new Scotland but its methods are anathema to him. He wants an independent Scottish Kingdom–and complete separation from England. To achieve this he will abduct the King.'

36

'Dugall suspected as much. Why did you not tell me earlier?' asked Luke.

'David and I are allies. On one occasion at Castle Clarke I entered his chamber unexpectedly and saw an amethyst ring on a small table. I was well aware of its significance. It is only recently that I came to doubt David's unconditional loyalty to the King. David has always been a strong nationalist. However I do not think he organized the explosion at Edinburgh Castle. I think personal hatred by Malcolm and Fenella provided the momentum for that incident. David came to that reception as a matter of protocol, and was probably warned by Fenella to leave the room just before the explosion. How were you stupid enough to have all your officers in the one room?'

'We were not. None of them were officers. Fenella's ploy of a reception was too obvious. We were prepared.'

'Though not well enough,' muttered Alistair. 'Despite his treachery to the King David saved our lives. If you were really aware of the plot why did you not warn us of the pending explosion?'

'I was outsmarted. I thought I had removed any danger of an explosion so there was nothing to warn you about,' explained a contrite Luke.

As both men pondered on the implications of their conclusions regarding David and the King there was a general hubbub throughout the castle. The guards were being replaced. Alistair recognised some of the new officers. These men were from Colonel Burns' own company. One of the new officers spied them half hidden behind a column not far from the King's chamber. He spoke firmly,

'Gentlemen there is a threat against His Majesty and we have been ordered to clear all people from the palace precinct. Please move across the courtyard.'

Both men had reached the same conclusion. The conspiracy against the King was underway and their old friend and comrade David Burns was its leader. Luke was blunt. 'David now controls all of the troops in the vicinity of Stirling. We need an overwhelming force to counter this dominance, and none is available.'

'This explains recent developments,' confessed Alistair. 'David sent most newly formed regiments north to liaise with the Highlander army near Inverness. In essence he has removed any regiments that might oppose his coup. You are probably well versed in military coups, Colonel. Is this not the way the English army acts against its political enemies?'

Luke smiled, 'The positive is that David will not harm the King. He will be placed under house arrest and govern on the advice of David and his friends. It is an old Scottish custom, as you know. The King's grandfather, and your great uncle, James, spent most of his life when King of Scotland as the houseguest of one political faction or the other. History is simply repeating itself.'

The two men left the castle, and walked through the grounds that contained the barracks of the newly recruited regiments which were loyal to Colonel David Burns. The night was clear and both men walked some distance into the foothills. Alistair suddenly stopped, 'Do you hear that?'

Luke answered, 'Bagpipes, a fair distance away.'

'Yes, let's hope that David, deep inside the castle, has not heard it. Let's find the source,' announced Alistair. 'Depending on its size and commander that Highland regiment might yet save the King.'

A Highland regiment had been moving down a major road to the west of Stirling, but with the onset of dusk it made camp for the night in the foothills just outside the city. The bagpipes ceased. After making their way up an ever-narrowing track, Luke and Alistair were eventually challenged in Gaelic. Alistair replied in kind and demanded to know which regiment was moving so close to Stirling. Both men breathed a sigh of relief. They were

Cameron highlanders under Sir James Cameron with several companies of Irish Royalists that had evaded General Monk's blockade.

Sir James was delighted to see Alistair, but remained suspicious of an English officer to whom he had never warmed. Alistair explained the situation. Sir James was aghast. 'The King is a prisoner?'

'As a Scot you would say he is a guest of Colonel Burns, and in the morning the King will announce a new policy that will conform with the views of Burns and his supporters,' explained Alistair. 'The King will be forced to declare that he no longer intends to invade England.'

Luke intervened, 'At the moment no one knows that the King's freedom is being restricted. The King himself may not yet know that he is a prisoner. If we could change the situation before David makes the position public the King's credibility will be preserved.'

'Why would a republican Englishman be concerned with the welfare of our King?' asked the pragmatic Sir James. 'Your new ally should be Colonel Burns, not our King whom you despise.'

'True, I am not concerned about the so-called King but firstly Burns lied to me for months, and recently did not warn me of an attempt on my life. Secondly General Cromwell wants Charles to invade England so that we can put an end to his pretensions once and all.'

The Highlanders were ready to march within the hour. Under the cover of darkness they took over the Stirling garrison unbeknown to those in the King's chamber. At first light Alistair and Luke sought an audience with the King. After waiting outside the royal chamber for over an hour they were admitted to the royal quarters having relinquished their swords and pistols to the guards. The King lay on his bed clearly distraught. David had already revealed his hand.

The royal retinue was gone, and David sat in one of the large chairs normally reserved for the King. Dugall was slumped against the wall. His nose was bloodied, and his weapons had been removed.

As Luke and Alistair entered the King jumped to his feet,

'Alistair, are you part of this outrage?'

'No, sire we are here to rescue you.'

David laughed, 'Two men against the cream of Scotland's new army. The old Earl of Barr's regiment has been recreated and controls not only the castle, but also the environs of Stirling. I sent all other regiments away from our temporary capital. The King will today order all the units from the Highlanders in the north through our various regiments between here and there to move on the English. A nationalist Scottish movement will drive you out of our country in the name of Charles, King of Scotland.'

Alistair and Luke made their way across the room and stood beside the King. Alistair spoke quietly but firmly, 'Your Majesty, we will escort you from the room.'

David was amused. 'There are ten fully armed troopers in this room. I could probably take the two of you myself without seeking their help, especially as you have been disarmed. Alistair you are under arrest, and Luke you will leave immediately for Edinburgh, or you will hang as a spy.'

Luke whispered something to Charles. Alistair sat against the wall next to Dugall. Both men were disconsolate. Luke suddenly grabbed the King and dragged him into the middle of the room. He placed a dagger, which he had hidden on his person, at the lad's throat. There was general sigh of concern. David was suddenly jolted out of his satisfied reverie. His men looked at him for instructions.

'You English cur. What do you hope to achieve?' he exclaimed.

'Everything. If I kill the lad your entire plan falls apart. The boy's brother the next alleged King is not in your hands. When the truth is known you will be blamed for this King's death.'

'What do you want Luke?' asked David tersely.

'Order your men out of the room. Rearm Dugall, Alistair and myself. I will then release the so-called King into Dugall's custody.'

David smiled to himself. There was no way out for his opponents. He could afford to humour them and later easily regain the upper hand. When the time was right he would simply summon back his men and annihilate his opponents. His troops left the room. Luke, Alistair and Dugall regained their weapons. Dugall, with his pistol, dirk, and sword restored took his task as personal protector of the King very seriously. David allowed his antagonists to play out their little game. Then he decided to test Luke's position and conviction.

'Luke I do not know why you have adopted the position you have. It is certainly not in England's interest. In the name of the King I can offer you a deal. If you can persuade your government to withdraw all its troops from Scotland the King will guarantee that he will not invade England, or take any steps to reclaim his English throne. That is a deal that you can ill afford not to place before your government.'

'David, if you really were in the position to dictate such terms I would certainly take them to my government, but they are just empty words. Your attempted coup has failed.'

Luke had heard a bagpipe, the signal that Sir James's men were in position. He then announced, 'Dugall, escort your King out of the chamber.'

'What good will that serve Luke? 'asked Dugall. 'Colonel Burns's men are outside.'

'Trust me Dugall, the men outside the door will not harm their King.'

Luke moved to the door, opened it slightly, and half pushed Dugall and Charles out into the corridor. He quickly closed the door. David sarcastically commented, 'What are you hoping to achieve Luke? By now my men will have disarmed Dugall, and placed the King under their protection.'

Alistair turned to David, 'Surrender your weapons to us and throw yourself on the King's mercy, or you will be dead by dawn.'

David had had enough of this farce. 'Which of you wants to be the hero?'

Alistair did not give Luke any time to respond. Alistair drew his sword and rushed at David. They parried for some time with Alistair the quicker, and David the stronger. The latter began to thrust more effectively with a sequence of attacks penetrating Alistair's doublet. Blood began to flow and after the next forceful parry David smashed Alistair's sword from his grasp and placed his against his victim's throat.

Luke intervened. 'Spare his life, and let him leave the room for someone to staunch his bleeding.' David nodded, and Luke helped Alistair to the door.

Luke turned towards David with his own sword drawn and asked, 'You are a professional soldier. Why have you betrayed your last two masters, the Earl of Barr and now Charles Stuart?'

'I have betrayed nobody. My masters are the traitors. They betrayed the survival of Scottish independence. I have been fighting for the same objective all my life. Now with the King in my hands he will rule as King of Scotland advancing the policies that I support. Initially that is to unite all the Scottish armies and then drive all the English forces from Scotland.

We will not waste our depleted resources on an invasion of England. In fact I will give you a letter for Cromwell, as I suggested, promising that if the English withdraw their troops immediately, the King will guarantee that the Scots will support the English Republic.'

'Let us settle this man to man,' said Luke waving his sword over his head.

'No Luke, I will not give you the satisfaction. I will not throw away my vision for Scotland by risking my life against one of England's best swordsmen.' He called loudly, 'Lieutenant, return with your men and arrest Colonel Tremayne.' He moved to the door to repeat his order when it burst open and the young King strode in followed by dozens of Highlanders.

David was stunned. The King ordered his immediate arrest. Sir James Cameron personally disarmed him, and with a detachment of Highlanders led him away. Charles was somewhat shaken. He thanked Luke and Alistair for their part in the rescue, as his entourage came flooding back into the room fawning over his person.

Luke returned to his room in the castle and as he had been awake all night was soon in a deep sleep. Some hours later he was forcibly awakened. Alistair, heavily bandaged, was shaking him furiously.

'David has escaped!' Alistair explained that a unit of David's regiment had remained concealed in the lower reaches of the castle and had not been found by the Highlanders. They overpowered the guards and released David from his cell, but they could not free themselves from the tower block which the Highlanders had now surrounded.'

When Alistair and Luke reached the courtyard Sir James informed them that the news was not so bad. David and his men were still confined to the small tower. Sir James turned to Luke,

'You English are experts at siege warfare. What do you recommend I do?'

'You have excellent artillery. Demolish the tower with Colonel Burns and men in it.'

'There is a problem with that. The tower is one of the castle's arsenals. If we shell it many of the buildings in this part of the castle would be demolished by the massive explosion that would follow. What is really worrying is that Burns could create the same result himself by igniting a few barrels of gunpowder. I have asked the King to leave the castle in case Burns wants to take us all with him.'

'But why would he do that? It would do nothing to advance his cause,' said Alistair.

'Sir James, David sees himself as a patriot. He cannot conceive of himself as a traitor. The King of Scotland would have no more fervent a supporter provided he did not weaken that position by assuming the Kingship of England. The easiest approach is to wait. Starve them out!' advised Luke.

'Again that is problem. Part of the arsenal is a storehouse for the castle's food supplies,' replied Sir James.

'You not going to tell me that it also has its own water supply that you cannot disrupt?' asked a cynical Luke.

'I do not like your tone Colonel Tremayne. Given this weather the men in that tower could gather enough snow and rainwater to survive a long siege. This is a stalemate.'

37

The discussion ended as David appeared at the highest window of the tower. He shouted to the officers gathered below, 'I have a proposition to put to the King. I wish him to renounce his authority to try me, and place the responsibility on a higher power. I place my trust in God and demand trial by the sword. I will duel with the King's representative and if I win, I will be freed.'

'Do not be silly man. Such a trial has been replaced by the King's law for centuries,' shouted back an irate Sir James.

Luke whispered, 'Agree to the proposition. It may be one way of getting him out of the tower.'

Sir James continued, 'Nevertheless it is my duty to pass your request on to the King.' He spoke to a few courtiers and then disappeared inside the palace precinct. Alistair followed him. They emerged some time later. Alistair was beaming, Sir James sullen.

The latter called up to David, 'The King rejects your formal request for trial by sword, but agrees that if you fight Dugall Sinclair–and win–he will agree to your exile, but the fight must be in the courtyard.'

David was bickering with Sir James over the detail of the engagement when a trumpet sounded. Charles entered the courtyard with a large entourage. David continued to shout attempting to gain the King's attention. Charles moved close to the base of the tower. This forced David to lean further out of the window to talk to him. Before anyone could react the air was thick with arrows flying from several directions. The King's French crossbowmen had fired from the courtyard at David as he leant forward to speak to the King. A second group of bowmen had fired from a neighbouring and higher tower down onto David's men gathered on the roof. David, taken by surprise and fatally wounded, fell to the cobbled courtyard below. Charles walked over to the crumpled body and removed an amethyst ring from the finger of the shattered corpse.

Before he left Stirling Luke had a drink with Alistair. Luke promised him that he would ensure that Elspeth was taken immediately to her new protector, the Marquis of Argyle. They hugged and formally saluted each other, and then walked away in opposite directions. They knew confrontation between the King's Scottish armies and the English invaders was imminent.

Luke returned to Edinburgh where Andrew brought him up to date on the bombing. He had disconnected the trail of gunpowder from the main door of the cellar to the barrels, but he had failed to see that there was another access to the chamber through a small trapdoor at the other end of the room. Someone had opened the trap door and dropped a lighted taper into an open container of gunpowder. It was a suicidal act. A number of large diamonds were found with some melted gold on bits of charred skeleton among the debris. The man who had lit the fatal explosion was a bejewelled Sir Malcolm Petrie. He would have been the first to confront the blast.

Luke kept his promise to Alistair and farewelled Elspeth who left immediately for the lands of the Marquis of Argyle to be reunited with her child. Luke smiled to himself when rumours reached Edinburgh claiming that the commander of the King's garrison at Stirling, a Colonel David Burns had had a fatal accident–falling from one of the castle's towers. Sir James Cameron was promoted to Major General and appointed to command all the King's troops. He had immediately sent couriers to recall all regiments in Scotland to Stirling. It was also reported that a gang of brigands had been found hidden within the town. They were identified as the same group that had killed two English noblemen. The Cameron Highlanders executed them by decapitation but their leader Duff Mackail was not accounted for.

Luke did not tell Donna of her brother's escape. The Scottish authorities would execute her for witchcraft the following day. Luke had wanted her tried for the murder of the Earl of Barr, but as a conviction for a capital crime was already recorded, it was deemed unnecessary. As a concession to her status she would not burn at the stake, but be hanged at noon. Donna walked from her cell to the gibbet that had been erected just outside the castle gate. The English garrison provided a small guard to protect her from the abuse of the gathering crowd of sightseers. Luke walked with Donna from the castle and down the slope. He helped her mount the stairs. Two masked hangmen were going about their business. One of them placed a hood over Donna's head.

Luke did not want to stay. He turned, and headed back towards the castle. But something was amiss. Why were there two hooded hangmen? Then he noticed a piebald pony tied to the gibbet. He ran back and saw Donna and the smaller of the hangmen in deep conversation. Luke had no time to think. Then he saw the hangman's boots. They had been shortened. He primed his pistol, placed it to the ear of the unsuspecting executioner and fired. The assembled crowd, at first stunned, began to scream abuse and catcall. Luke pulled off the mask of the shot hangman. There was enough of his face left. It was Duff Mackail. Luke immediately signalled the real hangman to dispatch Donna. A hushed audience burst into cheering and applause as the trapdoor swung open.

Two days later Luke visited Fenella who had been convicted by an English military court of the murder of several Scottish nobles and a number of unnamed English soldiers. Fenella was relaxed and outgoing, despite her imminent execution. Luke was surprised at her good cheer. She was anxious to talk and enjoyed retracing recent events. 'Luke, it has been an exciting few years. When Montstone raped me I told another visitor to the house, David Burns. Some time later he informed me that my own father had created a secret organization for the King that could be used to exact my revenge. When I was evicted from Montstone's house and my betrothal quashed, father decided to exact his revenge on the Barrs. I was delighted to find David the joint leader of your mission, and even more surprised when David revealed himself as a double agent. The King had appointed him leader of the Black Thistle in place of father. We agreed to destroy the Barr family, rally wavering Scottish nobles to the King, and remove the English military command as a symbol of our nationalism.'

'How did David react when the King abolished the Black Thistle?'

'David who had spent most of his working life at the upper levels of an independent Scottish government was very worried that the young King would be confused by the temptation of being returned to his English throne. He would sacrifice Scottish independence in the process. David believed we should use the Black Thistle to maintain Scottish independence by obliterating the English high command as well as any English royalist leaders who might poison the young King's mind. Towards the end he lost interest in the assassination of Barr supporters and the English military. He withdrew his support. David wanted all our resources redirected against Royalists who supported the King's English enterprise. He eventually decided these aims could best be realised by controlling the person of the King, as had been a Scottish custom.'

'Why did he involve us? Why did he convince Barr that the English needed to participate?'

'He knew of your so-called secret unit. He assumed that if Cromwell's life were threatened you would be called in. By working closely with you he hoped to wangle his way into the higher levels of the English military which would create opportunities for him to kill the General. It worked perfectly. You invited a whole cell of assassins to Edinburgh. How did you uncover the bomb plot?'

'Your very invitation immediately aroused our suspicion. Your insistence on holding the reception in that particular chamber well away from the kitchens confirmed our fears. A search of the cellars revealed the gunpowder. Let's go back a bit. How did you feel about people arriving at Castle Clarke that were not part of your plan?'

'Most of them were added by other members of the Black Thistle for their own personal reasons. Duff persuaded Mackelvie to stay, not because he left me for another woman but because he was a sleeping member of Black Thistle who refused to become active when he was needed. Duff did not trust Mackelvie. Young Morag was also kidnapped by Duff because he heard that she was regaining her memory regarding the murder of her parents, a fact which I innocently passed on to him. Duff was one of the murderers, and feared he might be recognized. I did my best to protect the girl. Duff was guilty of dozens of capital crimes so to be charged with the murder of the Ritchies would have made little difference. After his failed attempt to strangle her my view prevailed.'

Fenella continued, 'Alistair's arrival initially perplexed us, as he claimed to be a member of the Black Thistle sent by the King to dissolve the organization. It was then that David decided to use the organization for our own interests. The arrival of Mungo Macdonald was a major worry. David vaguely remembered the man as an agent of the Earl of Barr. We all thought that its enemies had infiltrated the Black Thistle until Macdonald's behaviour suggested he had a limited mission, to protect Elspeth. It was the conflict between Alistair and the Earl of Barr over Elspeth's baby that diverted you from the main game.'

'Why did David go through the charade of the murder in the tavern and the fight between assailants and ourselves in Oban, and the finding of a note incriminating the Black Thistle?'

'No great mystery. Although he was a double agent–leading the Earl of Barr's elite special force, and head of a secret society out to destroy the Earl– his men were loyal to the Earl. His deputy was a professional who would have detected David's deception if he had not played his role as the Earl of Barr's man to perfection. Duff's men carried out the decapitation and the faked assault on David and yourself in Oban. In the process he achieved a bonus of removing Barr's leading agent, Angus Mackelvie. David knew that Angus had been sent by Barr to supervise his activities.'

'What drove a minor laird such as David to assume such visions of grandeur for an independent Scotland?'

'David was more important than his current status suggested. He was not David Burns. Nor was he of lowly status. Fifty years ago the King outlawed one of the great clans of Scotland. The name of the clan could never be spoken and all members of it had to take another name. David would have been chief of this outlawed clan whose power and resources had been taken over by the Crown and shared among other clans. David hoped that his association with Barr would have led the government to render the act of outlawry null and void, and restore his clan to existence, and himself to prominence. His marriage to the Earl of Barr's niece was covert recognition of his higher status by the nobleman. When Barr failed to recreate the clan largely due to the opposition of the Marquis of Argyle who had benefited by its abolition David turned to the King. The chance encounter after the meeting of the Black Thistle at the woodcutter's hut gave David the chance to put himself at the service of the King. He believed a royal promise that the outlawing of his clan would be revoked. The King was grateful as David had saved his life, and insisted that David continue in his position with the Earl, while accepting the leadership of the Black Thistle.'

'You seem to know a lot about David's intimate history.'

'Yes, my mother's mother and David's grandfather were brother and sister.'

'And members of the outlawed clan?'

'Yes. David was consistent. In the twists and turns of Scottish politics he was one of the few men who stayed loyal to his principle, the maintenance of an independent Scotland. His final solution, controlling the person of the King, had many precedents in Scottish history.'

Fenella embraced Luke. As he climbed up the stairs from the dungeon cell a tear ran down his face. Fenella would hang at dawn. He spent the night with Gillian, and did not leave her bed until well after noon.

HISTORICAL EPILOGUE

In August 1651 the English defeated the Scots yet again, and Cromwell continued mopping up operations. Charles Stuart took advantage of this and with his depleted army marched into England, hoping to raise troops on his march south. Cromwell eventually surrounded the Royalists at Worcester where in September 1651 Cromwell gained another momentous victory. Charles after many adventures escaped to the continent where he would remain in exile for nine years. The English captured the remnant of the Scottish government, and most of Scotland. Major General George Monk became the effective English military ruler of Scotland. He marched south almost a decade later to place the exiled King on the throne of England.

OTHER LUKE TREMAYNE ADVENTURES

THE SPANISH RELATION

A gripping murder mystery set in 1655 against the potential collapse of the Cromwellian regime. Luke Tremayne is sent by Cromwell to investigate the death of a Somerset squire. Tremayne discovers that the squire's family and village are deeply divided, providing a multitude of suspects—feuding relatives, corrupt politicians, a secret Royalist society, clandestine militias, Spanish agents, religious fanatics and wanton witches. Luke stumbles on a plot to alter the government, and in the process experiences the suspicion, violence, bawdiness and superstitions of village life; enjoys the attention of several lusty women; and is perplexed by a man in a golden cape. (Trafford 2007)

THE IRISH FIASCO

In 1648 Cromwell sent Luke to Ireland to investigate the murder of a fellow officer; and the disappearance of a fortune in Spanish silver. Disaster strikes before he leaves England, and an obsessed deputy creates additional tensions. Luke is thwarted by the mysterious Arabella and confronted by several fanatics. Lust leads him to near death in the Wicklows where he discovers a secret Catholic refuge and a sensual witch. He uncovers a traitor and contributes to the military victory that paved the way for Cromwell's reduction of Ireland. In a final twist Luke's world is turned upside down. (Trafford 2008)

CHESAPEAKE CHAOS

In 1650 Luke is sent to Maryland on a mission for proprietor the Catholic Lord Baltimore and the Puritan general Oliver Cromwell, to assess the former's collapsing authority, and the situation should the English Republic seek to invade the Royalist colonies on the Chesapeake. Luke becomes involved with a dysfunctional planter family and quickly has several murders to solve, motives for which could stem from the current fraught situation, or from an atrocity committed during the Civil War. The situation is complicated by planter rivalry, Indian wars, a Puritan settlement and the incursion of an unfriendly foreign power. Luke struggles to solve the murders, and save the Chesapeake for England.

(TRAFFORD 2009)

Forthcoming

The Angelic Assassin

In 1652 with its foreign enemies subdued the English army directs its attention to its master – the remnant of the English Parliament, known as the Rump. As the corrupt politicians plot against their own army Luke is sent to investigate suspicious behaviour and the murder of government officials on the outskirts of an expanding London. Luke struggles against the breakdown of law and order and the unpopularity of both Rump and army. He uncovers a network of criminals and conspirators whose links with soldiers and politicians could change the course of history.

www.ingramcontent.com/pod-product-compliance
Lightning Source LLC
Chambersburg PA
CBHW030922120626
46554CB00001B/236